D0171783

# Biochemistry I

*By Frank Schmidt*

IDG Books Worldwide, Inc.
An International Data Group Company
Foster City, CA ♦ Chicago, IL ♦ Indianapolis, IN ♦ New York, NY

*About the Author*

Frank Schmidt, Ph.D., is Professor of Biochemistry at the University of Missouri-Columbia. Since 1978, he has taught biochemistry to graduate, medical, undergraduate, and continuing education students. His scholarly work is in the areas of RNA biochemistry, the origin of life, drug discovery, and inquiry-based science education.

*Publisher's Acknowledgments*

*Editorial*

Project Editor: Kathleen A. Dobie

Acquisitions Editor: Kris Fulkerson

Technical Editor: Cristina Furdui

Editorial Assistant: Laura Jefferson

*Production*

Proofreader: Melissa Buddendec

IDG Books Indianapolis Production Department

CLIFFSQUICKREVIEW™ Biochemistry I

Published by

IDG Books Worldwide, Inc.

An International Data Group Company

919 E. Hillsdale Blvd.

Suite 400

Foster City, CA 94404

www.idgbooks.com (IDG Books Worldwide Web site)

www.cliffsnotes.com (CliffsNotes Web site)

Library of Congress Control Number: 00-103374

ISBN: 0-7645-8563-0

Printed in the United States of America

10 9 8 7 6 5 4 3 2 1

1V/RS/QX/QQ/IN

Distributed in the United States by IDG Books Worldwide, Inc.

Distributed by CDG Books Canada Inc. for Canada; by Transworld Publishers Limited in the United Kingdom; by IDG Norge Books for Norway; by IDG Sweden Books for Sweden; by IDG Books Australia Publishing Corporation Pty. Ltd. for Australia and New Zealand; by TransQuest Publishers Pte Ltd. for Singapore, Malaysia, Thailand, Indonesia, and Hong Kong; by Gotop Information Inc. for Taiwan; by ICG Muse, Inc. for Japan; by Intersoft for South Africa; by Eyrolles for France; by International Thomson Publishing for Germany, Austria and Switzerland; by Distribuidora Cuspide for Argentina; by LR International for Brazil; by Galileo Libros for Chile; by Ediciones ZETA S.C.R. Ltda. for Peru; by WS Computer Publishing Corporation, Inc., for the Philippines; by Contemporanea de Ediciones for Venezuela; by Express Computer Distributors for the Caribbean and West Indies; by Micronesia Media Distributor, Inc. for Micronesia; by Chips Computadoras S.A. de C.V. for Mexico; by Editorial Norma de Panama S.A. for Panama; by American Bookshops for Finland.

For general information on IDG Books Worldwide's books in the U.S., please call our Consumer Customer Service department at **800-762-2974.** For reseller information, including discounts and premium sales, please call our Reseller Customer Service department at **800-434-3422.**

For information on where to purchase IDG Books Worldwide's books outside the U.S., please contact our International Sales department at **317-596-5530** or fax **317-572-4002.**

For consumer information on foreign language translations, please contact our Customer Service department at **800-434-3422,** fax 317-572-4002, or e-mail rights@idgbooks.com.

For information on licensing foreign or domestic rights, please phone **650-653-7098.**

For sales inquiries and special prices for bulk quantities, please contact our Order Services department at **800-434-3422** or write to the address above.

For information on using IDG Books Worldwide's books in the classroom or for ordering examination copies, please contact our Educational Sales department at **800-434-2086** or fax **317-572-4005.**

For press review copies, author interviews, or other publicity information, please contact our Public Relations department at **650-653-7000** or fax **650-653-7500.**

For authorization to photocopy items for corporate, personal, or educational use, please contact Copyright Clearance Center, 222 Rosewood Drive, Danvers, MA 01923, or fax **978-750-4470.**

# CONTENTS

# CONTENTS

# CONTENTS

# CONTENTS

# CHAPTER 1
## THE SCOPE OF BIOCHEMISTRY

*Biochemists discuss chemistry with biologists, and biology with chemists, thereby confusing both groups. Among themselves, they talk about baseball.* –Anonymous

As the name indicates, **biochemistry** is a hybrid science: Biology is the science of living organisms and chemistry is the science of atoms and molecules, so biochemistry is the science of the atoms and molecules in living organisms. Its domain encompasses all the living world with the unifying interest in the chemical structures and reactions that occur in living systems. Where can you find biochemistry? All through science, medicine, and agriculture.

Biochemistry underlies ordinary life in unseen ways: For example, take a middle-aged man (not very different from the author of this book) who:

- Takes a drug to lower his serum cholesterol. That drug was developed by a pharmaceutical company's biochemists to inhibit a key enzyme involved in cholesterol biosynthesis.

- Shaves with a cream containing compounds that soften his beard. These active agents were developed after studies of the physical properties of keratin, the protein in hair.

- Eats a breakfast cereal fortified with vitamins identified through nutritional biochemistry.

- Wears a shirt made from pest-resistant cotton. The cotton plants were bioengineered by biochemists through the transfer of genes from a bacterium into plants.

- Goes fishing after work. The conservation agents who manage the stream use biochemical information from the DNA (deoxyribonucleic acid) sequences to track the genetics of the fish population.

■ Drinks milk before bedtime. His sleep is helped by the amino acids in the milk, which are converted by his brain into molecular signals that lead to a resting state in other parts of his brain.

All these everyday events depend on an understanding of the chemistry of living systems. The purpose of this book is to provide a quick review of the chemical structures and events that govern so much of daily life. You can use it as a supplement to existing texts and as a review of biochemistry for standardized examinations.

## Biochemistry is a Contemporary Science

In the early nineteenth century, as chemistry became recognized as a scientific discipline, a distinction was made between inorganic and organic chemistry. Organic compounds (those containing carbon and hydrogen) were thought to be made only in living systems. However, in 1828, Friedrich Wöhler in Germany heated an inorganic compound, ammonium carbamate, and made an organic one, urea, found naturally in animal urine. Wöhler's experiment showed that the chemistries of the living and nonliving worlds are continuous:

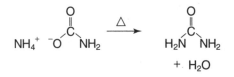

At the end of the nineteenth century, a parallel controversy arose as organic chemists debated whether an intact, living cell was needed to carry out biochemical reactions. Hans Büchner in Germany reproduced the synthesis of ethanol with a cell-free extract of brewer's yeast, showing that reactions of living systems can be reproduced *in vitro* (literally, in glass), that is, away from a living system. Reactions in living cells occur because they are catalyzed by enzymes — the very word *enzyme* is derived from the Greek word for yeast, *zymos*.

Biochemistry became a distinct science in the early twentieth century. In the United States, it arose from the merger of physiological chemistry and agricultural chemistry. Contemporary biochemistry has three main branches:

- **Metabolism** is the study of the conversion of biological molecules, especially small molecules, from one to another — for example, the conversion of sugar into carbon dioxide and water, or the conversion of fats into cholesterol. Metabolic biochemists are particularly interested in the individual enzyme-catalyzed steps of an overall sequence of reactions (called a pathway) that leads from one substance to another.

- **Structural Biochemistry** is the study of how molecules in living cells work chemically. For example, structural biochemists try to determine how the three-dimensional structure of an enzyme contributes to its ability to catalyze a single metabolic reaction.

- **Molecular Genetics** is concerned with the expression of genetic information and the way in which this information contributes to the regulation of cellular functions.

These distinctions are somewhat artificial, as contemporary biochemistry is intimately connected with other branches of biology and chemistry, especially organic and physical chemistry, physiology, microbiology, genetics, and cell biology.

Extrapolating Biochemical Information

If the reactions of every organism were different, biochemistry would be a poor science. Contemporary biochemistry depends on the ability to extrapolate information from one system to another. For example, if humans and animals made cholesterol in fundamentally different ways, scientists would have no way to find a compound to treat high cholesterol and prevent heart attacks. It would be impossible (and unethical) to screen the millions of known organic compounds in humans to find an effective treatment. On the other hand, using enzyme systems, researchers can screen many thousands of compounds for their ability to inhibit an enzyme system *in vitro.* They can then screen the small number of active compounds for their effectiveness in laboratory animals, and then in humans.

## Common Themes in Biochemistry

At first glance, the subject matter of biochemistry seems too complicated to do anything other than blindly memorize it. Fortunately, biochemistry has a number of unifying themes, which can help you keep the varying branches in perspective. Some of the themes you'll encounter repeatedly in this book are explored in the following sections.

### Biochemical reactions involve small molecular structures

Four classes of small molecules combine to make up most of the important bimolecular structures (see Figure 1-1). Most of these are *optically active,* that is, they are found in only one of the possible **stereoisomers.** **(Stereoisomers** are compounds that have the same kinds and numbers of atoms but have different molecular arrangements.

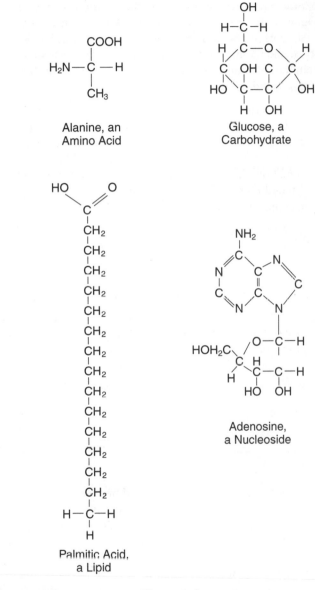

Alanine, an
Amino Acid

Glucose, a
Carbohydrate

Adenosine,
a Nucleoside

Palmitic Acid,
a Lipid

Figure 1-1

- **Amino acids** all have the common core structure shown in Figure 1-1. Generally, amino acids found in nature are the L-stereoisomers. Amino acids are the building blocks of proteins, and have an important role in energy metabolism and in cellular signaling. They are also a small but important part of cell membranes.

- **Carbohydrates** are molecules of the empirical formula $C_n(H_2O)_n$ where $n$ usually ranges from 3–7. They are found in sugars and starches and make up parts of nucleotides (the energy currency of a cell, and the building blocks for genetic information). They are also present in some components of all cell membranes. They are the central components of energy-producing pathways in biology.

- **Lipids** are closely related to hydrocarbons (compounds containing hydrogen and carbon atoms exclusively), although they usually have other atoms beside C and H. Characterized by limited solubility in water, lipids are essential components of membranes, and are important energy stores in plants and animals.

- **Nucleosides** and **nucleotides** contain a carbohydrate component joined to one of four carbon- and nitrogen-containing ring compounds called bases. They make up the energy currency of the cell, and, when joined end-to-end (polymerized) into DNA or RNA chains, form the genetic information of a cell.

Polymers in Living Systems

In the cell, single amino acids, sugars, and nucleotides can be joined together into **polymers.** Polymers are large molecules composed of small subunits arranged in a "head to tail" fashion. Living systems are based on polymers. There are several reasons why this is true:

- **Economy of synthesis:** Chemical reactions occur much more quickly and specifically in living cells than they do in an organic chemical reaction. The speed and specificity of biochemical reactions are due to the enzymes that *catalyze* the reactions in a cell. How does the cell get the many catalysts needed to support life? They can be made one by one or mass-produced. Mass production is much more efficient, as can be seen by the following exercise.

Suppose that a living system needs 100 catalysts. These catalysts could be synthesized one by one. Where would the catalysts to make the catalysts come from? Making the set of 100 catalysts would require at least 100 more catalysts to synthesize them, which would require 100 more catalysts, and so on. A living cell would need a huge number of catalysts, greater than the number of known organic molecules (or even the number of atoms in the universe). Suppose, on the other hand, that the catalysts were mass-produced. Joining the amino acids to each other by a common mechanism allows a single catalyst to join 20 different amino acids by the same chemical reactions. If two amino acids join together, they can make $20 \times 20 = 400$ possible **dimers** (molecules composed of two similar subunits); joining three together makes $20 \times 20 \times 20 = 8,000$ **trimers** (molecules made of three similar subunits), and so on. Because a single protein may contain 1,000 or more amino acids joined end to end, a huge number of different catalysts can be made from the relatively few monomer compounds.

- **Economy of reactions:** Joining monomers to make macromolecules is economical if the monomers can be joined by the same chemistry. If the monomers contained different functional groups, synthesis of each polymer would require a different kind of catalyst for each monomer added to the chain. Clearly, it is more economical to use a generic catalyst to put together each of the many monomers required for synthesis.

- **Stability of cells:** This argument is based on the properties of water. If red blood cells are placed in distilled water, they burst. Water moves across the membrane from the outside to the inside. In general, water moves across a membrane from the side with a lower solute concentration to the side with higher solute concentration; the side with higher solute concentration has a higher osmotic pressure. The cell has to expend energy to maintain its osmotic pressure. The osmotic pressure of a system is based on the number of atoms or molecules dissolved in water, not on their size. Thus, 100 molecules of a carbohydrate monomer (a sugar) have the same osmotic pressure as 100 polysaccharide molecules, each containing 100 monomers; however, the latter macromolecule can store 100 times more energy.

## Cell membranes

Organisms have an inside and an outside, and the reactants and products of biochemical reactions are kept either in the cell for further reaction or excreted. Membranes are formed by lipid bilayers as shown in Figure 1-2, with proteins dissolved in the lipid. The outsides of bilayers are charged and interact with water, while the insides are hydrocarbon-rich. Besides defining the boundaries of a cell, membranes are used for energy generation and for separating the components from each other when necessary. For example, powerful digestive enzymes in eukaryotic cells are kept inside a membrane-bounded compartment, the **lysosome.** Substrates for these enzymes are imported into the lysosome.

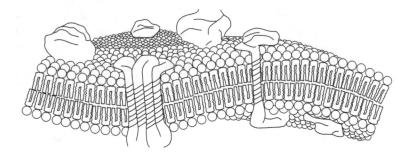

Figure 1-2

## Types of Biochemical Reactions

Although there are many possible biochemical reactions, they fall into only a few types to consider:

- **Oxidation and reduction:** For example, the interconversion of an alcohol and an aldehyde.

- **Movement of functional groups within or between molecules:** For example, the transfer of phosphate groups from one oxygen to another.

- **Addition and removal of water:** For example, hydrolysis of an amide linkage to an amine and a carboxyl group.

- **Bond-breaking reactions:** For example, carbon-carbon bond breakage.

The complexity of life results, not from many different types of reactions, but rather from these simple reactions occurring in many different situations. Thus, for example, water can be added to a carbon-carbon double bond as a step in the breakdown of many different compounds, including sugars, lipids, and amino acids.

### Regulating biochemical reactions

Mixing gasoline and oxygen can run your car engine, or cause an explosion. The difference in the two cases depends on restricting the flow of gasoline. In the case of the car engine, you control the amount of gasoline entering the combustion chamber with your foot on the accelerator. Like that process, it's important that biochemical reactions not go too fast or too slowly, and that the right reactions occur when they are needed to keep the cell functioning.

## Large molecules provide cell information

The ultimate basis for controlling biochemical reactions is the genetic information stored in the cell's DNA. This information is expressed in a regulated fashion, so that the enzymes responsible for carrying out the cell's chemical reactions are released in response to the needs of the cell for energy production, replication, and so forth. The information is composed of long sequences of subunits, where each subunit is one of the four nucleotides that make up the nucleic acid.

## Weak interactions and structural stability

Heat often destroys a biochemical system. Cooking a slice of liver at temperatures only slightly over 100°F. destroys the enzymatic activity. This isn't enough heat to break a covalent bond, so why aren't these enzymes more robust? The answer is that enzymatic activity and structure depend on weak interactions whose individual energy is much less than that of a covalent bond. The stability of biological structures depends on the *sum* of all these weak interactions.

## Biochemical reactions occur in a downhill fashion

Life on earth ultimately depends on nonliving energy sources. The most obvious of these is the sun, whose energy is captured here on Earth by **photosynthesis** (the use of the light energy to carry out the synthesis of biochemicals especially sugars). Another source of energy is the makeup of the Earth itself. Microorganisms living in deep water, the soil, and other environments without sunlight can derive their energy from **chemosynthesis,** the oxidation and reduction of inorganic molecules to yield biological energy.

The goal of these energy-storing processes is the production of carbon-containing organic compounds, whose carbon is reduced (more electron-rich) than carbon in $CO_2$. Energy-yielding metabolic processes oxidize the reduced carbon, yielding energy in the process. The organic compounds from these processes are synthesized into complex structures, again using energy. The sum total of these processes is the use of the original energy source, that is, light from the sun, for the maintenance and replication of living organisms, for example, humans.

The energy available from these reactions is always less than the amount of energy put into them. This is another way of saying that living systems obey the **Second Law of Thermodynamics,** which states that spontaneous reactions run "downhill," with an increase in **entropy,** or disorder, of the system. (For example, glucose, which contains six carbons joined together, is more ordered than are six molecules of $CO_2$, the product of its metabolic breakdown.)

## All Organisms are Related

The classification and grouping of organisms, the science called **taxonomy,** regards organisms as similar based on their visible characteristics. Thus, from the Greeks until recently, plants and animals were regarded as the two main kingdoms of life. Later, cell biologists divided organisms into **prokaryotes** and **eukaryotes,** that is, organisms without and with a nucleus. Most recently, a new taxonomy has been developed, largely by Carl Woese and associates, based on the information in the ribosomal RNA sequences. Ribosomal RNA, used as an evolution clock, is essential to life, easy to identify, and full of surprises.

Remarkably, the most information-rich classifications of life shows three main divisions, sometimes called **domains,** which are more fundamental than the distinction between plants and animals, or prokaryotes and eukaryotes. These domains are:

- **Eukarya:** The most familiar domain, eukarya includes organisms with a nucleus. This division includes plants, animals, and a large number of what are sometimes called **protists,** or organisms that can be seen only under a microscope, such as yeasts or paramecia.

- **Bacteria:** The second domain includes microorganisms without a nucleus, including many that are familiar, like *Escherichia coli.*

- **Archaea:** The third group, on the molecular/biochemical level, is as different from bacteria as they are from eukarya. These remarkable microorganisms inhabit niches often thought of as inhospitable to life — for example, locations with high temperature, low oxygen, or high salt. Their biochemistry is unique and largely unexplored. Fully half of the known genes of these organisms are apparently unique, with no counterparts in bacterial or eukaryotic genomes.

## The Common Origin of Organisms

The basis of the study of molecular evolution and taxonomy is the origin of organisms. Although the tree of life shown in Figure 1-3 was derived from sequences of a single gene, the similarities among organisms' biochemical and molecular properties are greatest for the organisms closely branched on the tree. Thus, human metabolism is more similar to that of chimpanzees, a close relative, than to that of yeast, a more distant relative. Human and yeast biochemistry are more similar to each other than either is to an archael or bacterial organism. The implications of this are important for the application of biochemistry to human disease. It is obviously unethical to do many biochemical experiments on human beings; however, animals or cultured animal cells are similar enough to find principles in common. For example, medical researchers can study the properties of genes that cause disease in mice, and evaluate potential treatments for safety before trying them on humans. Although not foolproof, this principle of similarity has been used continually in biomedical research dedicated to disease treatment and prevention.

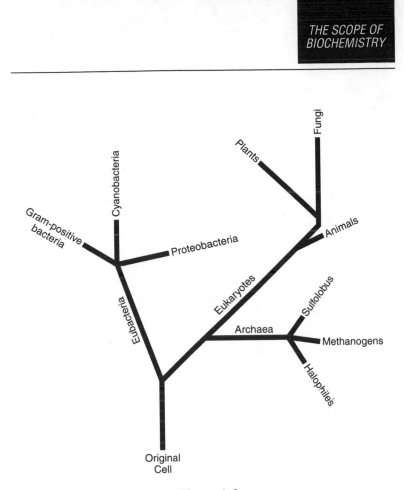

Figure 1-3

*Water, water everywhere.* –Coleridge

## The United Strength of Biochemical Structures

The forces that hold biomolecules together in three dimensions are small, on the order of a few kJ/mole, and much weaker than a covalent bond (formed through sharing of electrons between two atoms), which has an energy of formation a hundred times larger. Would life be possible if these molecules were held together only by covalent bonds? Probably not. For example, muscle contraction involves movement of the protein myosin relative to a filament composed of another protein, actin. This movement does not involve the breakage or formation of covalent bonds in the protein. A single contraction cycle requires about 60 kJ/mole; which is about 3% to -5% of the energy captured during the complete combustion of a mole of glucose. If the energy required for contraction were the same as that of forming a carbon-carbon covalent bond, almost all the energy of combustion of a molecule of glucose would be required for a single contraction. This would place a much higher demand for energy on the cell, which would require a similarly high demand for food on an organism.

If the forces holding them together are so small, how can biomolecules have any sort of stable structure? Because these small forces are **summed** over the entire molecule. For example, consider a double-stranded DNA a thousand base pairs long. The energy of an average base pair, about 0.5 kJ/mole, is not great, but the energy of 1,000 base pairs equals 500 kJ/mole, equivalent to the energy of several covalent bonds. This also has important consequences for the **dynamics** of individual base pairs: They can be opened easily while the molecule as a whole is held together. This property of weak interactions will become important in the consideration of DNA replication and transcription, later in this series.

## Properties of Water and Biomolecular Structure

Water is necessary for life. Many plant and animal adaptations conserve water — the thick skin of desert cacti and the intricate structure of the mammalian kidney are just two examples. Planetary scientists look for evidence of liquid water when speculating about the possibility of life on other planets such as Mars or Jupiter's moon, Titan.

Water has many remarkable properties, including:

- **High surface tension:** Despite being denser than water, small objects, such as aquatic insects, can stay on top of water surface.

- **High boiling point:** Relative to its molecular weight, water boils at a high temperature. For example, ammonia, with a molecular weight of almost 17, boils at -33° C, while water, with a molecular weight of 18, boils at 100° C.

- **Density is dependent on temperature:** Solid water (ice) is less dense than liquid water. This property means that lakes and ponds freeze from the top down, a benefit to the fish living there, who can overwinter without being frozen solid.

### The properties of water and hydrogen bonds

Water has a **dipole,** that is, a separation of partial electrical charge along the molecule. Two of oxygen's six outer-shell electrons form covalent bonds with the hydrogen. The other four electrons are nonbonding and form two pairs. These pairs are a focus of the partial negative charge, and the hydrogen atoms correspondingly become partially positively charged. Positive and negative charges attract each other, so that the oxygen and hydrogen atoms form **hydrogen bonds.** Each oxygen in a single molecule can form H-bonds with two hydrogens (because the oxygen atom has two pairs of nonbonding electrons). Figure 2-1 shows such a hydrogen bond. The resulting clusters of molecules give water its cohesiveness. In its liquid phase, the network of molecules is irregular, with distorted H-bonds. When water freezes, the H-bonds form the water molecules into a regular lattice with more room between the molecules than in liquid water; hence, ice is less dense than liquid water.

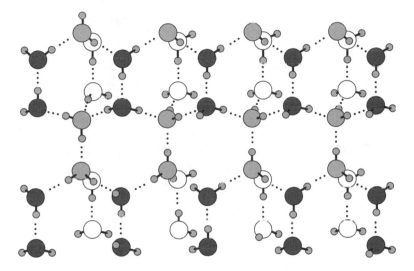

Figure 2-1

### Hydrogen bonds and biomolecules

In water, the nonbonding electrons are the **H-bond acceptors** and the hydrogen atoms are the **H-bond donors.** Biomolecules have H-bond acceptors and donors within them. Consider the side chain of a simple amino acid, serine. The oxygen contains two pairs of nonbonding electrons, as water does, and the hydrogen is correspondingly a focus of partial positive charge. Serine thus can be *both* an H-bond acceptor and donor, sometimes at the same time. As you would expect, serine is soluble in water by virtue of its ability to form H-bonds with the solvent around it. Serine on the inside of a protein, away from water, can form H-bonds with other amino acids; for example, it can serve as an H bond donor to the nonbonding electrons on the ring nitrogen of histidine, as shown in Figure 2-2.

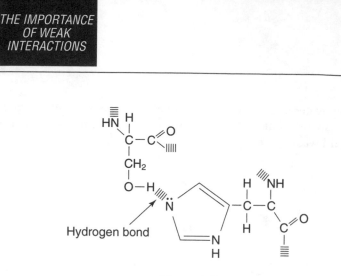

Figure 2-2

These H-bonds normally exist only when water is not present. If serine's side chain is found on the surface of a protein, it is very likely to form H-bonds, given the relatively high concentration of water available.

## The Hydrophobic Effect

### Nonpolar molecules and water-solubility

Because water is so good at forming hydrogen bonds with itself, it is most hospitable to molecules or ions that least disrupt its H-bonding network. Watching oils float on the surface of water demonstrates that oil molecules are nonpolar — they don't carry a charge or polarity, and do not dissolve in water. When an oil or other nonpolar compound encounters water, the compound disrupts the H-bonding network of water and forces it to re-form around the nonpolar molecule, making a cage of sorts around the nonpolar molecule. This cage is an ordered structure, and so is unfavored by the Second Law of Thermodynamics, which states that spontaneous reactions proceed with an increase in entropy (disorder).

How to resolve this dilemma? If the nonpolar molecules come together, then fewer water molecules are required to form a cage around them. As an analogy, consider the ordered water structure to be like paint around a cubic block. If you have four blocks to paint, and each block is 1 cm along each side, each block would require 6 cm$^2$ worth of paint if you paint them separately. However, if you put the four blocks together in a square pattern, you don't need to paint the inside surfaces of the cubes. A total of only 16 cm$^2$ rather than 24 cm$^2$ surface needs to be painted as Figure 2-3 shows.

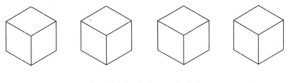

Each block has 6 sides
6 x 4 = 24

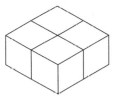

Four blocks together
have 16 sides exposed.

Figure 2-3

The tendency of nonpolar molecules to self-associate in water rather than to dissolve individually is called the **hydrophobic effect.** The term is somewhat misleading because it refers to the molecules themselves, where in reality it is due to the H-bonding nature of water, but it is used almost universally, and biochemists often speak of the hydrophobic side chains of a molecule as a shorthand for the complexities of discussing water structure as it is affected by nonpolar constituents of biomolecules.

Many biomolecules are **amphipathic,** that is, they have both hydrophobic (water-hating) and hydrophilic (water-loving) parts. For example, palmitic acid has a carboxylic acid functional group attached to a long hydrocarbon tail. When its sodium salt, sodium palmitate, is dissolved in water, the hydrocarbon tails associate due to the hydrophobic effect, leaving the carboxylate groups to associate with water. The fatty acid salt forms a **micelle** — a spherical droplet arranged with the hydrocarbon chains inside and the carboxylate groups inside on the outside of the droplet. Sodium palmitate is a major constituent of soap. Fats are **triglyceride esters,** composed of three fatty acids esterified to a single glycerol molecules. An ester linkage is a covalent bond between a carboxylic acid and an alcohol. Soap micelles mobilize fats and other hydrophobic substances by dissolving them in the interior of the micelle. Because the micelles are suspended in water, the fat is mobilized from the surface of the object being cleaned. Detergents are stronger cleaning agents than are soaps, mostly because their hydrophilic component is more highly charged than the fatty acid component of a soap. For example, sodium dodecyl sulfate is a component of commercially available hair shampoos. It is a powerful enough detergent that it is often used experimentally to disrupt the hydrophobic interactions that hold membranes together or that contribute to protein shape.

### Membrane associations

Glycerol esters of fatty acids are a large component of biological membranes. These molecules differ from those found in fats in that they contain only two fatty acid side chains and a third, hydrophilic component, making them amphipathic. Amphipathic molecules contain both polar (having a dipole) and nonpolar parts. For example, phosphatidylcholine, a common component of membranes, contains two fatty acids (the hydrophobic portion) and a phosphate ester of choline, itself a charged compound:

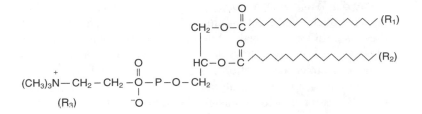

When phosphatidylcholine is suspended in water, the molecules associate by the hydrophobic effect, with the charged portion facing the solvent and the fatty acid side chains associating with each other. Instead of making a micelle, however, as palmitate does, these molecules associate into a **bilayer,** which eventually forms a spherical vesicle (termed a liposome) with a defined inside and outside. Liposomes are clearly similar to cell membranes, although they differ in some respects.

Biological membranes are bilayers and contain several types of lipids; some more often associated with the outside face of the cell, and others face the inside. Biological membranes also contain a large number of protein components. Membranes are *semipermeable,* naturally excluding hydrophilic compounds (carbohydrates, proteins, and ions, for example) while allowing oxygen, proteins, and water to pass freely.

### Electrostatic and van der Waals Interactions

Opposite charges attract. For example, $Mg^{2+}$ ions associate with the negatively charged phosphates of nucleotides and nucleic acids. Within proteins, salt bridges can form between nearby charged residues, for example, between a positively charged amino group and a negatively charged carboxylate ion. These electrostatic interactions make an especially large contribution to the folded structure of nucleic acids, because the monomers each carry a full negative charge.

**Van der Waals interactions** (see Figure 2-4) represent the attraction of the nuclei and electron clouds between different atoms. The nucleus is positively charged, while the electrons around it are negatively charged. When two atoms are brought close together, the nucleus of one atom attracts the electron cloud of the other, and vice versa. If the atoms are far apart (a few atomic radii away) from each other, the van der Waals force becomes insignificant, because the energy of the interaction varies with the 12$^{th}$ power of distance. If the atoms come closer together (so that their electron clouds overlap) the van der Waals force becomes repulsive, because the like charges of the nucleus and electron cloud repel each other. Thus, each interaction has a characteristic optimal distance. For two identical atoms, the optimal distance is d=2r, where r is atom radius. Within a biomolecule, these interactions fix the final three-dimensional shape. While van der Waals interactions individually are very weak, they become collectively important in determining biological structure and interactions.

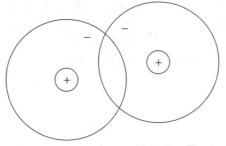

Negative Electron

Van der Waal's Interaction

Figure 2-4

## Acid-Base Reactions in Living Systems

Biochemists usually discuss acids and bases in terms of their ability to donate and accept protons; that is, they use the **Brønsted definition** of acids and bases. A few concepts from general chemistry are important to help organize your thoughts about biochemical acids and bases:

1. A compound has two components — a conjugate acid and a conjugate base. Thus, you can think of HCl as being composed of the proton-donating acidic part ($H^+$) and the proton-accepting basic part ($Cl^-$). Likewise, acetic acid is composed of $H^+$ and the conjugate base ($H_3CCOO^-$).

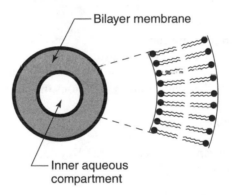

Bilayer membrane

Inner aqueous compartment

2. The stronger the acid, the weaker its conjugate base. Thus, HCl is a stronger acid than acetic acid, and acetate ion is a stronger base than chloride ion. That is, acetate is a better proton acceptor than is chloride ion.

3. The strongest acid that can exist in appreciable concentration in a solution is the conjugate acid of the solvent. The strongest base that can exist in a solution is the conjugate base of the solvent. In water, the strongest base that exists is $OH^-$. If a stronger base, such as $NaOCH_3$, is added to water, the methoxide ion rapidly removes protons from the solvent:

$$CH_3O^- + H_2O \rightarrow OH^- + CH_3OH$$

leaving the base $OH^-$ as the strongest base in solution. (Don't try these reactions at home; they are highly exergonic!) The strongest acid that can exist in water in appreciable amounts is $H_3O^+$, the conjugate acid of $H_2O$:

$$H_2O + H_2O \quad H_3O^+ + OH^-$$

4. Weak acids and bases — those less strong than $H^+$ or $OH^-$ — exist in equilibrium with water:

$$B + H_2O \quad BH + OH^-$$

$$HA + H_2O \quad H_3O^+ + A^-$$

## pK values and protonation

The strength of an acid or base is given by its $K_a$ or $K_b$, respectively. $K_a \times K_b = 10^{14}$, the dissociation constant of water. Just as it is convenient to describe the concentration of $H^+$ ions in solution as:

$$pH = -\log[H^+]$$

it is equally convenient to describe the $K_a$ of an acid as its negative logarithm, so that:

$$pK_a = -\log K_a$$

For example, acetic acid, which has a $K_a = 1.74 \times 10^{-5}$, has a $pK = 4.76$. Ammonia is more basic than water, with a $pK_a = 9.25$, corresponding to its $K_a = 5.6 \times 10^{-10}$. If the $pK_a$ of a group is < 7.0, it will donate a proton to water and a solution containing that compound will have a pH < 7, that is, it will be basic. Conversely, if the pK of a compound is > 7.0, that compound will accept a proton from water, and a solution containing that compound will have a pH > 7, which puts it in the acidic range.

## Solution pH

Many living organisms (there are many exceptions among the microbes) can exist only in a relatively narrow range of pH values. Thus, vegetables are often preserved by pickling them in vinegar, a dilute solution of acetic acid in water. The low pH of the solution prevents many bacteria and molds from growing on the food. Similarly, it is a cliche of movie Westerns that desert springs whose water is alkaline (basic) are decorated with the skulls of cattle who were unfortunate enough to drink from them. Finally, individuals with chest injuries who are unable to breathe efficiently develop a metabolic acidosis, as their blood pH drops below normal due to the impaired elimination of $CO_2$ (a weak acid) from the lungs.

Microorganisms capable of living in acidic environments expend a large amount of energy to keep protons from accumulating inside their membranes. These examples show the importance of controlling the pH of biological systems: Biochemical reactions, and therefore life, can exist only in a narrow, near-neutral pH range.

All physiological pH control relies ultimately on the behavior of *weak* acids and bases as buffers. A **buffer** is a combination of a weak acid and its salt or a weak base and its salt. The addition of an acid or a base to a buffered solution results in a lesser pH change than would occur if the acid were added to water alone.

This behavior is described quantitatively by the **Henderson-Hesselbach equation,** which can be derived from the definition of $K_a$:

$$K_a = [H^+] [A^-]/[HA]$$

where HA is a weak acid — acetic acid, for example.

Taking the logarithm of each side of the equation:

$$\log K_a = \log [H^+] + \log [A^-] - \log [HA]$$

Remembering that:

$$[A^-] - \log [HA] = \log([A^-]/[HA])$$

and multiplying through by (-1):

$$- \log K_a = -\log[H^+] - \log([A^-]/[HA])$$

Rearranging, and remembering the definitions of pK and pH:

$$\textbf{pH} = \textbf{pK}_a + \textbf{log } \textbf{([A}^-\textbf{]/[HA])}$$

This equation allows you to predict the pH of a buffered solution from the values of the $pK_a$ and the amount of basic and acidic forms of the buffer. (For convenience, the subscript of the pK brackets to indicate concentration and charges are sometimes omitted, and the equation becomes pH = pK +log (A/HA).)

For example, calculate the pH of a solution containing 0.1 M acetic acid and 0.01 M sodium acetate:

$$pH = 4.8 + \log (0.01/0.1) = 4.8 + \log (0.1) = 4.8 - 1 = 3.8$$

If the proportions of acid and salt are reversed, the pH would be 5.8. If they are equal, the ratio is:

$$[A]/[HA] = (0.01/0.01) = 1$$

and, because $\log 1 = 0$, $pH = pK_a = 4.8$.

**Buffer capacity**
What would happen if 0.005 equivalents of a strong acid, for example, HCl, were added to each of the preceding three solutions? The strong acid would donate protons to the acetate ion present in each solution. This would change the ratio [A]/[HA], and consequently, the pH, in each case:

**Case 1:** pH = 4.76 + log (0.005/0.105) =4.76 + (-1.32) = 3.4
instead of 3.8

**Case 2:** pH = 4.76 + log (0.095/0.015) = 4.76 + (0.80) =5.6
instead of 5.8

**Case 3:** pH = 4.76 + log (0.095/0.105) = 4.76 + (-0.04) = 4.7
instead of 4.8

If the HCl is added to pure water, the pH of the solution changes from 7 to 1.3. Thus, in each case, the change in pH was less than would have been observed in the absence of buffer. The lowest pH change is seen in case 3. This illustrates a general rule: *The amount of change in pH of a buffer system is lowest near the $pK_a$ of the conjugate acid.* In other words, buffers have their highest **capacity** when the amounts of acidic and basic components are nearly equal. In practice, buffers are generally useful when the ratio A/HA is between 0.1 and 10; that is, at a pH within ± 1 pH unit of their pKs.

### Biological acid-base equilibria

Metabolism occurs in cells at pH values near neutrality. For example, plasma must be maintained at a pH within half a pH unit of its normal value of 7.4. A number of mechanisms help accomplish this, including buffering by the mono- and di-basic forms of phosphate ion:

$H_3PO_4 + H_2O$ $H_2PO_4^- + H_3O^+$ pK = 2.14

$H_2PO_4^- + H_2O$ $HPO_4^{2-}$ $H_3O^+$ pK = 6.86

$H_2PO_4^{2-} + H_2O$ $PO_4^{3-} + H_3O^+$ pK = 12.4

At physiological pH, the Henderson-Hesselbach equation shows that the second equilibrium is most important. Phosphoric acid and phosphate exist in vanishingly small quantities near neutrality.

When carbon dioxide is dissolved in water, it exists in equilibrium with the hydrated form:

$$CO_2 + H_2O \ H_2CO_3$$

which is a weak acid. Carbonic acid, $H_2CO_3$, can donate two protons to a base:

$$H_2CO_3 + H_2O \ HCO_3^- + H_3O + pK_{a1} = 6.37$$

$$HCO_3^- + H_2O \ CO_3^{2-} + H_3O^+ \ pK_{a2} = 10.25$$

Metabolism releases $CO_2$, which reduces the pH of the fluid around the cell and must be buffered for metabolism to continue. In animals, hemoglobin and other blood proteins play an important role in this buffering.

*Cells obey the laws of chemistry.* –J.D. Watson

Consider the simple reaction of nitrogen to make ammonia:

$$N_2 + 3H_2 < \;\rightleftharpoons\; 2NH_3.$$

About half of the world's production of ammonia is carried out industrially and half biologically. At first glance, the two processes look quite different. The industrial reaction takes place at 500°C. and uses gaseous hydrogen and a metal catalyst under high pressure. The biological reaction takes place in the soil, uses bacterial or plant reactors, and occurs at moderate temperature and normal atmospheric pressure of nitrogen. These differences are so substantial that, historically, they were interpreted by supposing that biological systems are infused with a vital spirit that makes life possible. However, the biological reaction can be done with a purified enzyme. The biological reduction of nitrogen is more similar to than different from its industrial counterpart: The energy change from synthesis of a mole of ammonia is identical in both cases, the substrates are the same, and the detailed chemical reaction is similar whether the catalyst is a metal or the active site of an enzyme.

## Types of Metabolic Reactions

Metabolism refers to the dynamic changes of the molecules within a cell, especially those small molecules used as sources of energy and as precursors for the synthesis of proteins, lipids, and nucleic acids. These reactions occur in the **steady state** rather than all at once. *Steady state* refers to dynamic equilibrium, or **homeostasis,** where the individual molecules change but the rate at which they are made equals the rate at which they are destroyed. Concentrations of individual molecules in metabolic reactions are therefore kept relatively constant, while any individual molecules are present only for a brief time. Metabolism therefore is said to be an open chemical system. Metabolic reactions can be **catabolic** (directed toward the breakdown of larger molecules to produce energy), or **anabolic** (directed toward the energy-consuming synthesis of cellular components from smaller molecules).

## Enzyme Catalysts

Like almost all biochemical reactions, the biological synthesis of ammonia requires a specific biochemical catalyst—an **enzyme**—to succeed. Enzymes are usually proteins and usually act as true catalysts; they carry out their reactions many times.

## Space and Time Links in Metabolic Reactions

Thousands of distinct chemical reactions occur in a cell at any moment. A bacteria must simultaneously replicate its DNA, synthesize new enzymes, break down carbohydrates for energy, synthesize small components for protein and nucleic acid synthesis, and transport nutrients into and waste products out of the cell. Each of these processes is carried out by a series of enzymatic reactions called a **pathway.** The reactions of a pathway occur in succession, and the substrates for the pathways are often channeled through a specific set of enzymes without mixing. For example, in muscle cells, the glucose used to supply energy for contraction does not mix with the glucose used for transporting ions across the cell membrane.

## Energy Flow

Thermodynamics is the branch of chemistry and physics that deals with the energy flow in physical systems. The First Law of Thermodynamics states that energy in a system is neither created nor destroyed. The Second Law of Thermodynamics deals with the question of whether a reaction will occur: Spontaneous reactions occur with an increase in the **entropy** of a system; that is, the overall **disorder** of the system will increase. Entropy can be thought of as the energy that is not available to do work. For example, the oceans contain vast amounts of thermal energy in the form of the motions of individual water molecules. Yet this energy cannot be extracted to power a boat—that requires fuel or wind energy.

The amount of energy available for work is termed the **free energy** of a system and is defined as the difference in heat content between the products and reactants, less the amount of entropy change (multiplied by the temperature of the system):

$$\Delta G = \Delta H - T\Delta S$$

where $\Delta G$ is the amount of free energy released from the reaction, $\Delta H$ is the change in heat content, or **enthalpy,** T is the temperature in degrees Kelvin, and $\Delta S$ is the change in entropy. Another way of stating the Second Law of Thermodynamics is that reactions occur in the direction in which the free energy change is negative.

The change in free energy of a reaction in the standard state (conventionally, all reactants and products at 1M) is related to the equilibrium constant for the reaction by the following relation:

$$\Delta G° = - RT \ln(K_{eq}),$$

where R is the gas constant, T is the temperature in degrees Kelvin, and $K_{eq}$ is the equilibrium constant for the reaction. A reaction that is favored has $K_{eq} > 1$, and a reaction that is unfavored has $K_{eq} < 1$. In the former case, $\ln(K_{eq})$ is positive, so the free energy change of a favored reaction is negative; it is **exergonic.** Conversely, the natural logarithm of a number less than 1 is negative; because the product of two negative numbers is positive, the free energy change of an unfavored reaction is positive; it is **endergonic.**

Free energy changes associated with a biochemical reaction are determined at a **standard state,** with all reactants and products at 1M. Many biomolecules are unstable in acid, so the biochemical standard state is set at pH = 7.0 rather than at pH = 0 (1M acid), the standard state for chemical reactions. Biochemical standard free energies of reaction are given as $\Delta G^{\circ\prime}$ to where the ' indicates this change in standard conditions.

Biomolecules are also present in much lower concentrations than the standard state of 1M. The free energy change associated with reactions under conditions other than the standard state is given by the relationship:

$$\Delta G = \Delta G^{\circ} = + \text{ RT ln } (\Pi[\text{Products}]/\Pi[\text{Reactants}])$$

where $\Pi$ [Products] represents the concentrations of the products of the reaction, multiplied together, and $\Pi$ [Reactants] represents the concentrations of the reactants, multiplied together. (If more than one molecule of a product or reactant is involved in a reaction, the term for that component is raised to the power of the number of molecules involved in the reaction, just as for any chemical reaction.) R is the gas constant, and T is the absolute temperature of the reaction. This relationship reduces to $\Delta G = \Delta G^{\circ} =$ when all the products and reactants are present at 1M concentration, the standard state. At equilibrium, $\Delta G = 0$, and the equation reduces to:

$$\Delta G^{\circ\prime} = - RT \ln(K_{eq}).$$

$K_{eq}$ is simply the equilibrium constant of the reaction at pH 7.0.

## Free Energy Calculations

Biochemical free energies are usually given as standard free energies of hydrolysis. For example, the hydrolysis of glucose-6-phosphate:

glucose-6-phosphate + $H_2O$  glucose + $P_i$

has $\Delta G° = -4.0$ kcal/mole (-16.5 kJ/mole) under standard conditions. Therefore, the opposite reaction, the phosphorylation of glucose, is unfavored. However, the phosphorylation of glucose occurs readily in the cell, catalyzed by the enzyme hexokinase:

glucose + ATP  glucose-6-phosphate + ADP + phosphate

The other half of the phosphorylation reaction is the hydrolysis of ATP to yield ADP and inorganic phosphate ($P_i$):

ATP + $H_2O$  ADP + $P_i$

under standard conditions has $\Delta G° = -7.3$ kcal/mole ( -31 kJ/mole).

The standard free energy change of the reaction can be determined by adding the two free energies of reaction:

glucose + $P_I \rightarrow$ glucose-6-phosphate + $H_2O$ and
$\Delta G° = + 4.0$ kcal/mole

Note that the reaction as written is unfavored; its free energy change is positive. Another way of stating this is that the reaction is **endergonic,** that is, the reaction involves a gain of free energy.

For the **exergonic** hydrolysis of ATP (the reaction involves a loss of free energy):

ATP + H2O  ADP + Pi $\Delta G° = -7.3$ kcal/mole

The two reactions are summed:

glucose + ATP  glucose-6-phosphate + ADP + $P_i$ and

$\Delta G° = -3.3$ kcal/mole

This is a simple example of energetic **coupling,** where an unfavorable reaction is driven by a favorable one, as shown in Figure 3-1.

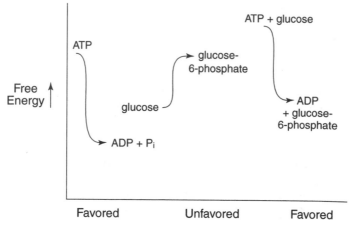

Figure 3-1

Coupling doesn't occur all by itself. In this example, if this experiment were set up so that the ATP would have to be hydrolyzed in one tube and the glucose phosphorylated in another, no coupling would be possible. Coupling can occur only when the partial reactions are part of a larger system. In this example, coupling occurs because both partial reactions are carried out by the enzyme hexokinase. In other cases, coupling can involve membrane transport, transfer of electrons by a common intermediate, or other processes. Another way of stating this principle is that coupled reactions must have some component in common.

## The Cell's Energy Currency

As noted in the previous section, the hydrolysis of ATP to yield ADP and phosphate is highly exergonic. This loss of free energy is due to the structure of the phosphoanhydride, which involves two negatively charged groups being brought into close proximity. Additionally, the phosphate group is stabilized by resonance not available to the anhydride (see Figure 3-2).

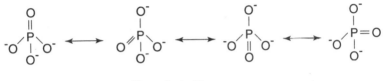

Phosphate Resonance

Figure 3-2

Because the free energy of hydrolysis of ATP's first two phosphates is so highly negative, biochemists often use the shorthand term **high energy phosphate** to describe the role of ATP in the cell. In general, the reactions of catabolism lead to the **synthesis** of ATP from ADP and phosphate. Anabolic reactions, as well as the other reactions involved in cellular maintenance, use the coupled **hydrolysis** of ATP to drive the reactions. For example, a muscle fiber will metabolize glucose to synthesize ATP. The ATP can be used to drive muscle contraction, to synthesize proteins, or to pump $Ca^{2+}$ ions out of the intracellular space. ATP serves as cellular energy currency because it is a common component of many reactions. It serves this role so well because it is **metastable:** In the cell, it does not break down extensively by itself over time (kinetic stability), but at the same time, it releases large amounts of free energy when it is hydrolyzed to release inorganic phosphate (thermodynamic instability).

All the free energy calculations shown in the previous examples have been done in the standard state, with all the products and reactants present at 1M concentration. However, very few compounds, except perhaps water, are present in the standard state. Because the free energy change of a reaction under nonstandard concentration is dependent on the concentrations of products and reactants, the actual $\Delta G$ of the reaction of glucose and ATP will be given by the equation:

$$\Delta G' = \Delta G^{\circ'} + RT \ln ([\text{glucose–6–phosphate}] [\text{ADP}][P_i]/[\text{glucose}][\text{ATP}])$$

The ratio of ATP to ADP is kept very high, greater than 10 to 1, so the actual $\Delta G$ of ATP hydrolysis is probably greater than 10 kcal/mole. This means that the reaction of ATP and glucose is even more favored than it would be under the standard state.

LeChatelier's Principle is fundamental to understanding these relationships. A reaction is favored if the concentration of reactants is high and the concentration of products is low. The free energy relationships shown in this section are a quantitative way of expressing this qualitative observation.

## Free-Energy-Driven Transport across Membranes

Cells expend a large amount of their free energy currency keeping the appropriate environment inside the cell. Thus, for example, $Ca^{2+}$ is present intracellularly at $< 10^{-7}$ M, while extracellular $Ca^{2+}$ is present in millimolar ($10^{-3}$ M) concentrations, that is, 10,000-fold higher. The free energy difference due to the difference in $[Ca^{2+}]$, sometimes termed its **chemical potential,** can be calculated. The difference of the $\Delta G^{\circ'}$ values when $Ca^{2+}$ is at the same concentration (1M) on each side of a membrane is, of course, zero, so the free energy is given by:

$$\Delta G = 0 + RT \ln ([Ca^{2+}]_{in}/[Ca^{2+}]_{out})$$

Converting from natural to base 10 logarithms, and substituting values for the gas constant, and a standard temperature of 25° C. (298° K.):

$$\Delta G = 2.303 \times 1.99 \text{ cal/°/mole} \times 298° \times (-4)$$

$$\Delta G = -5.5 \text{ kcal/mole}$$

This expression means that the influx of $Ca^{2+}$ into a cell is highly exergonic. If a channel is opened into a cell to allow $Ca^{2+}$ across the membrane, it will flood into the cell. In muscle cells, this influx of $Ca^{2+}$ is the signal for contraction. Cells, especially muscle cells, have a $Ca^{2+}$ active transport system, which transports two $Ca^{2+}$ ions out of the cell for every ATP hydrolyzed. The $\Delta G°'$ of ATP hydrolysis is enough to do transport only a single ion. Because, however, the ATP/ADP ratio is kept very high during active metabolic conditions, the concentration gradient of higher $[Ca^{2+}]$ outside the cell is maintained.

*One gene, one enzyme.* –George Beadle and Edward Tatum

## Complexity in Biochemical Genetics

At first glance, the subject of biochemical genetics can seem incomprehensibly complicated. How can a cell's genes possibly contain all the information about its capabilities for metabolism, macromolecular interactions, and responses to stimuli?

This question was answered, incorrectly, in the 1930s when biochemists concluded that the protein components of chromosomes had to carry genetic information. Scientists considered the DNA in chromosomes to be too simple a structure to be anything other than a scaffold. But in the 1940s, experiments carried out by Avery, Macleod, and McCarty showed that this view was wrong. Their experiments with bacteria showed that DNA carried the information for a heritable trait. This result forced a redefinition of the ideas about information in biology, and it was only when the Watson-Crick structure was proposed for DNA that it was understood how a "simple" molecule could carry information from one generation to the next. Although there are only four subunits in DNA, information is carried by the linear sequence of the subunits of the long DNA chain, just as the sequence of letters defines the information in a word of text.

The possible information contained in a biomolecule is termed its **complexity.** In molecular biology and biochemistry, complexity is defined as *the number of different sequences in a population of macromolecules.* Even a relatively small polymer has an enormous number of potential sequences. DNA, for example, is built from only four monomers: A, C, G, and T. If each of these monomers is linked with every other one, these 4 monomers now produce/contain 16 possible dimers ($4 \times 4$) because each position can have an A, C, G, or T. There are 64 possible trimers, $4 \times 4 \times 4$. So in any DNA chain the number of possible sequences is $4^N$, where N is the chain length.

Even a relatively small DNA chain can carry a large amount of information. For example, the DNA of a small virus, 5,000 nucleotides long, can have $4^{5,000}$ possible sequences. This is a huge number—approximately 1 with 3,010 zeroes after it. (By comparison, the number of elementary particles in the universe is estimated at $10^{80}$, or 1 with 80 zeroes after it.) But the virus has only one DNA sequence, which means that only one of the huge number of possible sequences has been selected to encode the virus's biochemical functions. In other words, there is *information* in the DNA sequence. The virus carries a large amount of information in a small space.

This concept of information is similar to the memory of a computer, which is made up of small semiconductor switches, each of which has two positions—on and off. The ability of computers to do an ever-increasing number of tasks depends on the ability of engineers to design chips that have more and more switches in a small space. Similarly, the ability of cells to do so many biochemical tasks depends on the large number of DNA nucleotides in the small space of the chromosomes.

## The Central Dogma of Molecular Biology: DNA Makes RNA Makes Protein

The sum total of all the DNA in an organism is called its **genome.** Genomic information is like a computer program for a cell. When you open a computer program, the program is copied from ROM (read-only memory) on the hard disk to RAM (random-access memory). The instructions in RAM are the ones that actually carry out the program, but the copy of the program in RAM exists only as long as there is power to the machine; if your PC loses power, you have to restart the program, and it is once again copied from the disk to RAM. This arrangement (hopefully) insures against the master copy of the program being damaged through a power surge or operator error.

If DNA is the master copy (the ROM) of a cell's genetic program, its integrity must be preserved. One way the DNA is protected is because **RNA** acts as the working copy (the RAM). Chemically, RNA is very similar to DNA. Biochemically, the major difference is that RNA either acts as a component of the metabolic machinery or is a copy of the information for protein synthesis. The relationship between DNA and RNA is called the **central dogma** of molecular biology:

### DNA makes RNA makes protein

In the first of these processes, DNA sequences are transcribed into **messenger RNA** (mRNA). Messenger RNA is then translated to specify the sequence of the protein. DNA is replicated when each strand of DNA specifies the sequence of its partner to make two *daughter* molecules from one parental double-stranded molecule.

### DNA, RNA, and nucleotide structure

DNA is a polymer—a very large molecule made up of smaller units of four components. Each monomer contains a **phosphate** and a **sugar** component. In DNA, the sugar is deoxyribose, and in RNA the sugar is ribose.

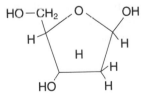

Deoxyribose

and one of four **bases**, two of which are **purines:**

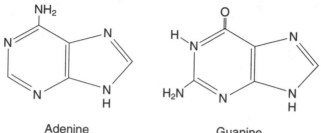

Adenine          Guanine

**Purine Bases**

and two of which are **pyrimidines:**

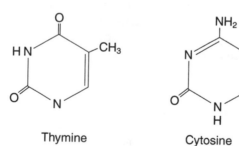

Thymine          Cytosine

**Pyrimidine Bases**

A sugar and a base make up a **nucleoside.** A base, sugar, and phosphate combine to form a **nucleotide,** as in thymidine monophosphate or adenosine monophosphate:

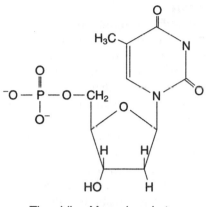

Thymidine Monophosphate,
a Deoxyribonucleotide

RNA is similar to DNA, although RNA nucleotides contain **ribose** rather than the deoxyribose found in DNA. Three bases found in DNA nucleotides are also found in RNA: adenine (A), guanine (G), and cytosine (C). Thymine in DNA is replaced by **uracil** in RNA:

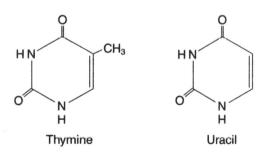

Thymine                     Uracil

**DNA's duplex nature**

DNA is normally **double-stranded.** The sequences of the two strands are related so that an A on one strand is matched by a T on the other strand; likewise, a G on one strand is matched by a C on the other strand. Thus, the fraction of bases in an organism's DNA that are A is equal to the fraction of bases that are T, and the fraction of bases that are G is equal to the fraction of bases that are C. For example, if one-third of the bases are A, one-third must be T, and because the amount of G equals the amount of C, one-sixth of the bases will be G and one-sixth will be C. The importance of this relationship, termed *Chargraff's rules,* was recognized by Watson and Crick, who proposed that the two strands form a **double helix** with the two strands arranged in an **antiparallel** fashion, interwound head-to-tail, as Figure 4-1 shows.

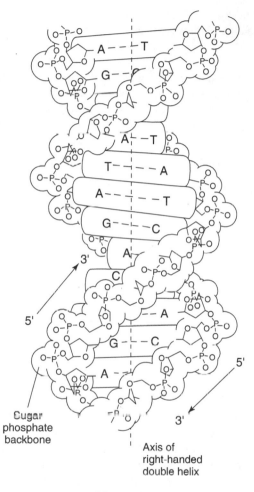

Figure 4-1

You usually read nucleic acid sequences of DNA in a **5' to 3'** direction, so a DNA dinucleotide of (5') adenosine-guanosine (3') is read as AG.

The **complementary** sequence is CT, because both sequences are read in the 5' to 3' direction. The terms 5' and 3' refer to the numbers of the carbons on the sugar portion of the nucleotide (the base is attached to the 1' carbon of the sugar).

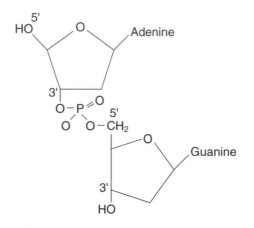

Complementarity is determined by **base pairing**—the formation of hydrogen bonds between two complementary strands of DNA. An A–T base pair forms two H-bonds, one between the amino group of A and the keto group of T and the second one between the ring nitrogen of A and the hydrogen on the ring nitrogen of T. A G–C base pair forms three H-bonds, one between the amino group of C and the keto group of G, one between the ring nitrogen of C and the hydrogen on the ring nitrogen of G, and a third between the amino group of G and the keto group of C. DNA's double helix is a result of the two strands winding together, stabilized by the formation of H–bonds, and of the bases **stacking** on each other, as Figure 4-2 shows.

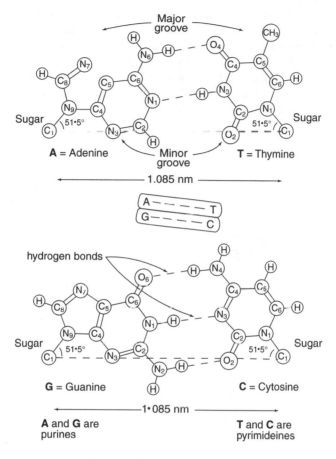

Figure 4-2

## The DNA double helix and genetic replication

Because an A on one strand must base-pair with a T on the other strand, if the two strands are separated, each single strand can specify the composition of its partner by acting as a **template.** The DNA template strand does not carry out any enzymatic reaction but simply allows the replication machinery (an enzyme) to synthesize the complementary strand correctly. This dual-template mechanism is termed **semi-conservative,** because each DNA after replication is composed of one parental and one newly synthesized strand. Because the two strands of the DNA double helix are interwound, they also must be separated by the replication machinery to allow synthesis of the new strand. Figure 4-3 shows this replication.

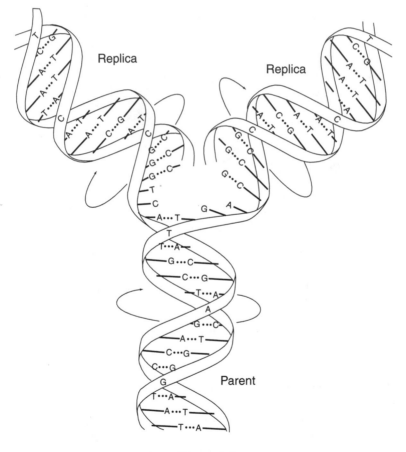

Figure 4-3

## RNA Carries Genetic Information

The two strands of DNA contain complementary information, so that one strand of DNA contains the information to specify the other strand. Normally, only one of the two DNA strands is copied to make RNA, in the process called **transcription.** RNA molecules, in contrast to DNA, are almost always single-stranded. **Base-pairing** determines the sequence of the RNA so that a DNA sequence (3')ATCCG(5') is copied into the RNA sequence (5')UAGGC(3').

Unlike DNA, RNA is disposable: Many copies of an RNA sequence are made from a single DNA sequence. These copies are used and recycled back to their constituent nucleotides. This allows the cell to respond quickly to changing conditions by transcribing different sequences into RNA. Special sequences called **promoters** tell **RNA polymerase,** the enzyme responsible for transcription, where to start making RNA (Figure 4-4).

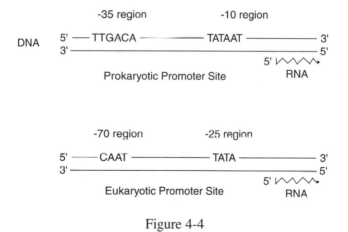

Figure 4-4

**Messenger RNA specifies the order of amino acids in proteins**

Proteins are linear polymers of amino acids. The sequence of a protein's constituent amino acids determines its biochemical function. The mRNA sequence is read in groups of three, called **codons.** Because there are four bases in DNA or RNA, there are 64 ($4^3$) codons. Only 20 amino acids are specified by translation, so there is more than one codon per amino acid. In other words, the genetic code is *redundant.* The code also contains punctuation marks. Three codons, UAG, UAA, and UGA, specify stop signals (like the periods in a sentence). One amino acid, methionine, coded by AUG, is used to initiate each protein (like a capital letter at the beginning of a sentence). Just as a letter that starts a sentence can also appear in an uncapitalized form inside the sentence, so methionine also appears internally in proteins. See Table 4-1.

Almost all organisms use the same genetic code. There are some differences, due primarily to the overall base composition of an organism's DNA. For example, *Mycoplasma* bacterial DNA is very high in A + T. Consequently, the TGG sequence (corresponding to the UGG codon) is rare, and the UGA codon specifies the amino acid tryptophan rather than a stop signal.

**Table 4-1: Genetic Code**

| First position (end) | Second position | | | | Third position (end) |
|---|---|---|---|---|---|
| | *U* | *C* | *A* | *G* | |
| | Phe | Ser | Tyr | Cys | U |
| U | Phe | Ser | Tyr | Cys | C |
| | Leu | Ser | STOP | STOP | A |
| | Leu | Ser | STOP | Trp | G |

| First position (end) | Second position | | | | Third position (end) |
|---|---|---|---|---|---|
| | **U** | **C** | **A** | **G** | |
| | Leu | Pro | His | Arg | U |
| C | Leu | Pro | His | Arg | C |
| | Leu | Pro | Gln | Arg | A |
| | Leu | Pro | Gln | Arg | G |
| | Ile | Thr | Asn | Ser | U |
| A | Ile | Thr | Asn | Ser | C |
| | Ile | Thr | Lys | Arg | A |
| | MET | Thr | Lys | Arg | G |
| | Val | Ala | Asp | Gly | U |
| G | Val | Ala | Asp | Gly | C |
| | Val | Ala | Glu | Gly | A |
| | Val | Ala | Glu | Gly | G |

The abbreviations for the amino acids are: phe, phenylalanine; leu, leucine; ile, isoleucine; met, methionine; val, valine; ser, serine; pro, proline; thr, threonine; ala, alanine; tyr, tyrosine; his, histidine; gln, glutamine; asn, asparagine; lys, lysine; asp, aspartic acid; glu, glutamic acid; cys, cysteine; trp, tryptophan; arg, arginine; gly, glycine.

**Transfer RNA**
Transfer RNA (tRNA) is the adaptor between mRNA and protein information. tRNA provides the specificity for the genetic code, so each codon doesn't have to specify a particular amino acid. Transfer RNA contains two active sites.

- The **anticodon** consists of three nucleotides that form base-pairs with the three nucleotides of a codon.

- The **acceptor** end is esterified to the amino acid specified by the codon.

The amino acid is loaded onto the acceptor end by an **aminoacyl–tRNA synthetase** enzyme (see Figure 4-5).

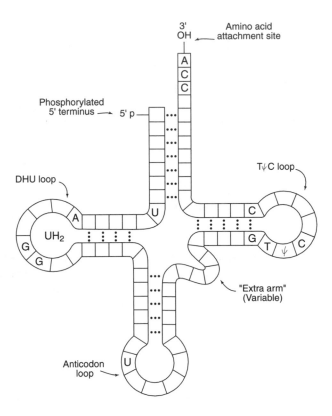

Figure 4-5

## Ribosomes and translation

**Ribosomes** are large particles composed of about two-thirds RNA and one-third protein by weight. Ribosomes facilitate several reactions:

- Initiation of the synthesis of a protein

- Base-pairing between the codon in mRNA and the anticodon in tRNA

- Synthesis of the peptide bond

- Movement of the mRNA along the ribosome

- Release of the completed protein from the translation machinery

Ribosomes consist of two subunits: a small subunit primarily involved with initiation, codon-anticodon interaction, and protein release; and a large subunit primarily concerned with the actual synthetic process:

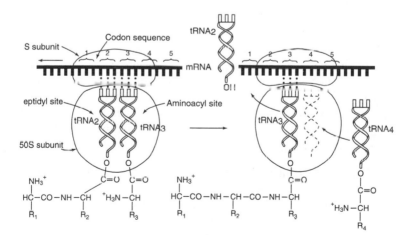

## Base-Pairing and the Central Dogma

All the interactions between nucleic acid molecules that help express genetic information involve base-pairing between *complementary* sequences. Complementarity is sometimes defined as selective stickiness. Complementary molecules fit together. In the case of nucleic acids, complementarity generally involves base pairing. For example, mRNA is complementary to one strand of DNA, and the anticodon of tRNA is complementary to the codon in mRNA. Replication, transcription, and translation all involve base-pairing at several levels.

### Genetic information expression

The central dogma allows the controlled expression of genetic information. Consider an *Escherichia coli* bacterium in its natural environment, the human gut. Its survival and replication would be favored by being able to use a variety of sugars to produce energy. On the other hand, making enzymes requires a large amount of energy. The conflict between these two demands is resolved by the bacterial genome's ability to synthesize the enzymes needed for digestion of sugars only when needed. Thus, for example, the enzymes involved in lactose digestion are made only when lactose is present in the environment. Usually, the synthesis of different proteins is controlled transcriptionally, that is, through regulating the synthesis of mRNA. When an *E. coli* bacterium encounters lactose, it synthesizes the mRNA species encoding the enzymes that degrade lactose. These mRNAs are translated into protein and the proteins catalyze the reactions required to digest lactose. After the mRNAs are translated they are degraded in the cell, so the control system contains the means of shutting itself off, as well.

This arrangement allows the *amplification* of DNA information. One DNA sequence, if transcribed into 20 mRNAs, each of which is translated into 20 protein molecules, can encode 400 ($20 \times 20$) enzymes, each of which can catalyze the breakdown of thousands of lactose molecules. All sorts of organisms use variations of this simple control model to control their growth and replication, the synthesis of macromolecular components, such as ribosomes, and a wide variety of anabolic and catabolic capabilities.

# CHAPTER 5
# PROTEIN STRUCTURE

## Levels of Protein Structure

The name protein comes from the Greek *proteios,* meaning primary. Although many other important biomolecules exist, the emphasis on protein as fundamental is appropriate. Proteins serve as important structural components of cells. More importantly, almost all the catalysts, or **enzymes,** in biological systems are composed of proteins. Proteins are linear chains of **amino acids** joined by **peptide bonds.** Twenty amino acids are incorporated into a protein by translation (for a reminder of these concepts, see Chapter 4). In some proteins, the amino acids are modified by subsequent post-translational events. The **sequence** of amino acids of a protein is termed its **primary structure.**

The amino acid chain, or backbone, forms one of a few **secondary structures,** based on the interactions of the peptide bond with nearby neighbors. The secondary structure that a chain forms is determined by the primary structure of the chain. Some amino acids favor one type of secondary structure, others prefer another, and still others are likely to form no particular secondary structure at all. Secondary structures are based on the interactions of closely neighboring amino acids.

The 20 amino acids differ in the nature of their **side chains,** the groups other than the repeating peptide unit. Interactions among the amino acid side chains within a single protein molecule determine the protein's **tertiary structure.** Tertiary structure is the most important of the structural levels in determining, for example, the enzymatic activity of a protein. Folding a protein into the correct tertiary structure is an important consideration in biotechnology. The usefulness of a cloned gene is often limited by the ability of biochemists to induce the translated protein product to assume the proper tertiary structure. (In the cell, specialized proteins, called **chaperonins,** help some proteins acheive their final structure.)

Finally, protein chains interact with each other as subunits associate to make a functional species. For example, hemoglobin, the mammalian oxygen carrier, contains two each of two different subunits. The ability of hemoglobin to deliver oxygen to the tissues is dependent on the association of these subunits. Interaction of proteins to form a **multimer** composed of several subunits is termed the protein's **quaternary structure.** Quaternary structure is often very important in determining the regulatory properties of a protein.

## Amino acids

The naturally occurring amino acids have a common structure. Amino acids, as the name implies, have two functional groups, an amino group (–NH₂) and a carboxyl group (–COOH). These groups are joined to a single (aliphatic) carbon. In organic chemistry, the carbon directly attached to a carboxyl group is the alpha ($\alpha$) position, so the amino acids in proteins are all **alpha-amino acids.** The side chains that distinguish one amino acid from another are attached to the alpha carbon, so the structures are often written as shown in Figure 5-1, where R stands for one of the 20 side chains:

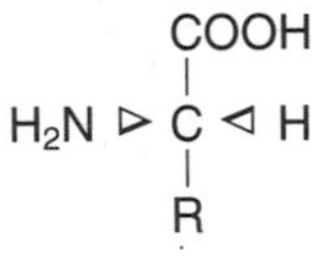

$$
\begin{array}{c}
\text{COOH} \\
| \\
\text{H}_2\text{N} \rhd \text{C} \lhd \text{H} \\
| \\
\text{R}
\end{array}
$$

Figure 5-1

The amino acids found in proteins have a common **stereochemistry.** In the structure illustrated in Figure 5-1, the amino group is always to the left side of the alpha carbon. In organic chemistry, this stereochemistry is referred to as **L** (for *levo,* meaning left). Thus, the amino acids found in proteins are L-alpha amino acids. (Biochemists, being creatures of habit, usually do not refer to amino acid stereochemistry in the R and S nomenclature.) A few D-amino acids are found in nature, although not in cellular proteins. (The D comes from *dextro,* meaning right.) For example, some peptide antibiotics, such as bacitracin, contain D-amino acids.

The carboxyll and amino groups of the amino acids can respectively donate a proton to and accept a proton from water. This exchange happens simultaneously in solution so that the amino acids form doubly ionized species, termed **zwitterions** (from German *zwei,* meaning two) in solution. The formation of zwitterions can be rationalized from the principles of acid-base chemistry discussed in Chapter 2. The strongest acid that can exist in water is the conjugate acid of water, the hydronium ion, $H_3O^+$. Carboxyllic acids are stronger acids than water, so the carboxyl group of an amino acid ($pK_a$ near 2) will donate a proton to water. Similarly, $\alpha$-amino groups ($pK_a$ greater than 9) are stronger bases than water and will accept a proton from water. Amino acids in water, therefore, have the general structure:

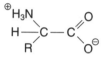

## Amino acid side chains

The side chains of amino acids give them their different chemical properties and allow proteins to have so many different structures. How many proteins are possible? Protein chains generally vary in size from 100 to 1,000 amino acids in length. Even if limited to the smallest chain length, there would be $20^{100}$, over $10^{130}$—that is, 1 with 130 zeroes after it—possible primary structures. (Again, remember that the number of elementary particles in the universe is estimated to be $10^{80}$.) Obviously, not all these potential proteins exist in nature. Instead, the primary structures of proteins are related to each other, and almost all proteins have **homologues,** that is, other proteins sharing a common ancestor.

What homologues are possible? In general, homologous proteins share some short amino acid sequences exactly. In other cases, the differences result in the substitution of one amino acid side chain by another chemically similar one. Six classes of amino acid side chains exist; within a group, the amino acid side chains are chemically similar. Substitution of one amino acid side chain for another one within the same group is known as **conservative** substitution. Homologous proteins are related by conservative amino acid substitutions, as in Figure 5-2. Although nonconservative substitutions are tolerated at some positions in the primary sequence of a protein, the general rule illustrated in Figure 5-2 is followed when evaluating the relationship of two protein primary sequences. (The dashes indicate that all three proteins have the same amino acid at that position—these are highly homologous proteins, indeed!)

```
DQFRDNAVRLMQSTPVIDGHNDLPWQLLKKFNNQLQDPRANLTSLNSTHT
-Q---L-V-I-QDT--------------NL---Q---PG---SS-AH---
-F---E-E-I-RDS--------------DM---R---ER---TT-AG---

NIPKLKAGFVGAQFWSAYTPCDTQNKDSVKRTVEQIDVIQRMCQLYPETF
-----K-----G----A-V------R-A-K--L--I--IQ---QA-----
-----R-----G----V-T------K-A-R--L--R--VH---RM-----

LCVTDSAGIQQAFQEGKVASLVGVEGGHSIDSSLGVLRALYHLGMRYLTL
AC--S-T-IR---R-------V-----------------H-----M--
LY--S-A-RR---R-------I-----------------Q-----L--

THSCNTPWADNWLVDTGEDKAQSQGLSSFGQSVVKEMNRLGIIIDLAHVS
----------------DDKA------H---S----M----VM-------
----------------DSEP------P---R----L----VL-------

VATMEAALQLSKAPVIFSHSSAYSLCHHRRNVPDHVLQLVKQTGSLVMVN
----R-A-K--Q-----------L-PH------D--Q---E-G------
----K-T-Q--R-----------V-AS------D--R---Q-D------

FYNDYVSCKAEANLSQVADHLDYIKKVAGAGAVGFGGDYDGVSRLPSGLE
---D-V--SAK-----------H--K------------Y---S-V-S---
---N-I--TNK-----------H--E------------F---P-V-E---

DVSKYPDLVAELLRRQWTEEEVRGALAENLLRVFKAVEQASDHKQAPGEE
--------V------Q---A--R----D------E------NHA-V-G--
--------I------N--A--K----D------E------NLT-A-E--

PIPLGGQLEASCRTKYGY SGTPSLHLQPGSLLASLVTLLLSLCLL    SL-D
----G--EA----N--- -AAP---LPP-S-----VP-L-LSLP    PK-RD
----D--GG----H---S-GAS---RHW-L-----AP-V-CLSLL    HK-MDP
```

Figure 5-2

**Aliphatic amino acids.** The side chains of glycine, alanine, valine, leucine, and isoleucine, shown in Figure 5-3, contain saturated carbon-carbon and carbon-hydrogen bonds only. Thinking of glycine as containing a side chain can be somewhat confusing because the fourth substituent on the $\alpha$-carbon is only a single hydrogen atom. Alone among the 20 amino acids, glycine is not optically active; the D- and L- nomenclature is irrelevant. Alanine has a methyl group for its side chain, valine a 3-carbon side chain, while leucine and isoleucine have 4-carbon side chains.

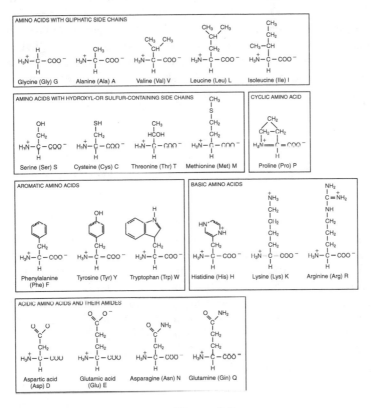

Figure 5-3

**Aromatic amino acids.** Phenylalanine, tyrosine, and tryptophan contain ring systems. In order of increasing complexity, phenylalanine has a benzyl group, while tyrosine is phenylalanine with an added hydroxyl group in the trans position relative to the methyl group. Tryptophan has two rings, one of which contains a nitrogen atom. The nitrogen is not ionizable at biologically relevant pH values.

**Ionizable basic amino acids.** Histidine, lysine, and arginine each have a nitrogen atom which, unlike the nitrogen of tryptophan, is ionized at the pH ranges found in the cell. Histidine has a 5-member imidazole ring. One of the two nitrogen ions has a $pK_a$ near 7.0. This means that, at the neutral pH values found in cells, about half of the histidine molecules will have their side chains protonated (that is, with a positive charge) and about half will have their side chains unprotonated and uncharged. Histidine is often used in enzymes to bind and release protons during the enzymatic reaction.

Lysine and arginine are almost fully ionized at the pH values found in the cell. Lysine's $pK_a$ is greater than 9; therefore, it will be > 99% protonated in the cell. Arginine's side chain is even more basic; its $pK_a$ is > 12. Therefore, these amino acids have a **net positive charge** in the cell.

**Carboxyllate-containing amino acids.** Aspartic acid and asparagine have four carbons; glutamic acid and glutamine have five carbons in all. Aspartic acid has a carboxyllic acid, and aspargine has an amide side chain. Similarly, glutamic acid has a carboxyllic acid side group, and glutamine has an amide group. The $pK_a$'s of the side chain carboxyll groups in aspartate and glutamate are near 4.0. Therefore, these side chain groups are almost fully ionized in the neutral conditions found in cells and are negatively charged.

## Hydroxyl and sulfur-containing side chains
**Serine and cysteine** can be thought of as being related to alanine. Serine is alanine with a hydroxyl (–OH) group and cysteine is alanine with a sulfhydryl (–SH) group.

**Threonine** has four carbons, with a hydroxyl group on the beta carbon. The beta carbon is next to the one containing the alpha carbon (the alpha carbon has the amino group on it). The presence of the hydroxyl group on threonine means that the beta carbon of threonine is optically active, in addition to the alpha carbon. As the name suggests, the –OH group has the D configuration, or threo to the alpha carbon. (The other possible stereochemistry is erythro—think of the letter E to remember this term. The arms of the E point in the same direction.)

**Methionine** has a *methyl* group on its sulfur. The backbone of methionine has one more carbon than does cysteine. (Cysteine with an extra carbon is termed **homocysteine;** homocysteine is an intermediate in the biosynthesis of methionine.)

### The cyclic amino acid
**Proline** is the odd one out among the amino acids. It has four carbons, with the alpha amino group bonded not only to the alpha carbon but also to the last side chain carbon. The cyclic side chain means that proline is conformationally rigid. That is, the carbon-carbon bonds of proline do not rotate in solution. Other amino acids are more flexible in solution.

**Peptide bond.** The peptide bond joins the carboxyl and amino groups of amino acids. When activated, carboxyllic acids and amines form *amides*. Amino acids are **bifunctional,** with each amino acid having both amino and carboxyl groups. Peptides are composed of amino acids joined head to tail with amide bonds. Peptides are classified according to their chain length. **Oligopeptides** are shorter than **polypeptides,** although no defined transition exists between the two forms. The joining of amino acids to form a peptide bond occurs formally (although the mechanism of its formation is more complicated) in the following way:

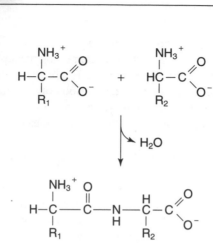

Note that two amino acids can form a **dipeptide** (a peptide composed of two units) in either of two ways. For example, glycine and alanine can form glycylalanine (gly-ala) or alanylglycine (ala-gly):

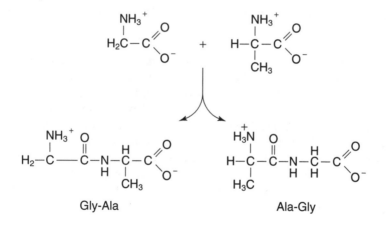

No matter which arrangement occurs, each dipeptide will have one free amino group and one free carboxyl group. Peptide sequences are written in the direction from the amino to the carboxyl end.

**Peptide bond structure.** The peptide bond structure favors coplanar N, C, and O atoms. Although a peptide bond is formally a carbon-nitrogen single bond, the unpaired electrons on the carboxyl oxygen and on the nitrogen can overlap through their pi orbitals to make the three-atom system partially double-bonded in character. The partially double-bonded system makes it harder to rotate the peptide bond in solution. As a result, peptide bonds can exist in one of two **conformational isomers,** with the two carbons either cis or trans to each other.

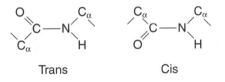

Usually, the trans conformation is favored. Proline is an exceptional case because its peptide bond does not have a hydrogen, making the cis and trans isomers harder to change one for the other:

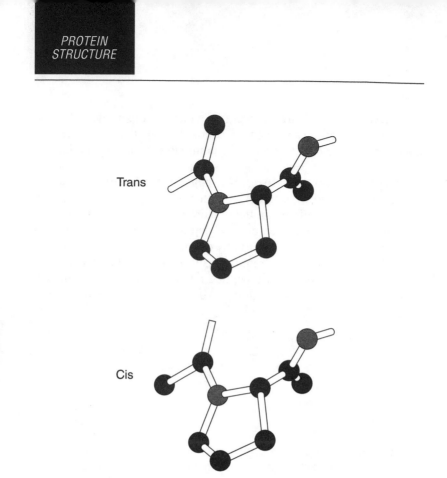

Trans

Cis

Proline usually is found in the trans isomer, although conversion (isomerization) between the cis and trans forms can be catalyzed by specific enzymes.

**Primary structure**

Protein sequences can be determined directly or from the DNA that encodes them. The sequence of amino acids in a protein determines its biological function. Direct determination of the amino acid sequence of an unknown protein is accomplished first by cutting the protein into smaller peptides at specific residues. For example, cyanogen bromide cleaves proteins after methionine residues, and the enzyme trypsin preferentially cleaves proteins after lysine and arginine residues. The amino acid sequences of the individual peptides are determined by removing the amino acids one by one through a set of reactions known as Edman degradation. This process leaves the question of how to order the different peptides, a step usually accomplished by comparing the sequences of the individual peptides arising from different cleavage steps.

The direct determination of an amino acid sequence is a very tedious process. Most protein sequences are now determined by determining the DNA sequence of the gene that encodes them; the amino terminus of a protein is the only portion that is determined directly. The sequence of DNA is then converted to the corresponding **predicted protein sequence** by using the genetic code to translate the codons into amino acids.

**Amino acid modification.** Although determining a DNA sequence is easier than identifying a protein sequence, the information obtained is incomplete and can be misleading. A large variety of functional groups may be added or removed from the side chains of amino acids in the protein. For example, two cysteine residues can be oxidized to form a **disulfide** bond:

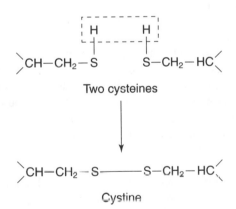

Two cysteines

Cystine

Sometimes the two joined cysteines are called **cystine.** Disulfide bond formation is an oxidative process; the sulfurs each lose a hydrogen atom, becoming more oxidized. Proteins found outside the cell are more likely to have disulfides than are proteins found inside the cell. This is due to the more reducing environment in the cell. Thus, for example, digestive enzymes found in the small intestine have disulfides, while many enzymes involved in cell metabolism have free (reduced) cysteine –SH groups.

Many other functional groups can be added to proteins post-translationally. Sugars or oligosaccharides can be added to the side chain oxygen of serine (O-linked glycosylation) or to the side chain nitrogen of asparagine (N-linked glycosylation). Phosphates can be added to the side chain oxygen atoms of serine, threonine, or tyrosine; this process is often important in cellular regulation. Proteins can be **covalently** bonded to each other. For example, the side chains of the protein collagen (found in skin and connective tissue) are linked together. These cross-linked residues are important in preserving tissue integrity, and, as humans age, they become more extensive, leading to the well-known loss of flexibility in old age.

### Secondary structure

The term *secondary structure* refers to the interaction of the hydrogen bond donor and acceptor residues of the repeating peptide unit. The two most important secondary structures of proteins, the alpha helix and the beta sheet, were predicted by the American chemist Linus Pauling in the early 1950s. Pauling and his associates recognized that folding of peptide chains, among other criteria, should preserve the bond angles and planar configuration of the peptide bond, as well as keep atoms from coming together so closely that they repelled each other through van der Waal's interactions. Finally, Pauling predicted that **hydrogen bonds** must be able to stabilize the folding of the peptide backbone. Two secondary structures, the **alpha helix** and the **beta pleated sheet,** fulfill these criteria well (see Figure 5-4). Pauling was correct in his prediction. Most defined secondary structures found in proteins are one or the other type.

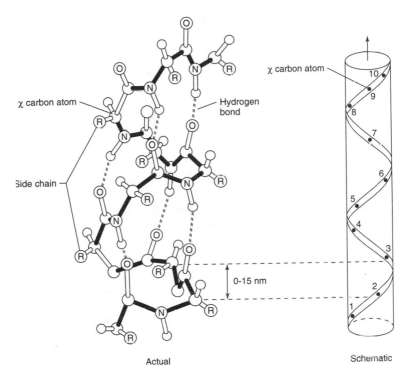

χ carbon atom

Hydrogen
bond

χ carbon atom

Side chain

0-15 nm

Actual

Schematic

Figure 5-4

**Alpha helix.** The alpha helix involves regularly spaced H-bonds between residues along a chain. The amide hydrogen and the carbonyl oxygen of a peptide bond are H-bond donors and acceptors respectively:

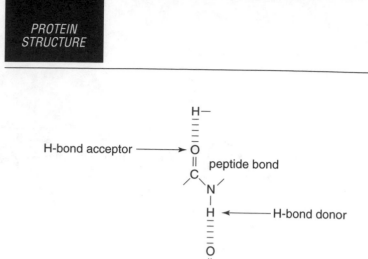

The alpha helix is **right-handed** when the chain is followed from the amino to the carboxyl direction. (The helical nomenclature is easily visualized by pointing the thumb of the right hand upwards—this is the amino to carboxyl direction of the helix. The helix then turns in the same direction as the fingers of the right hand curve.) As the helix turns, the carbonyl oxygens of the peptide bond point upwards toward the downward-facing amide protons, making the hydrogen bond. The R groups of the amino acids point outwards from the helix.

Helices are characterized by the number of residues per turn. In the alpha helix, there is not an integral number of amino acid residues per turn of the helix. There are 3.6 residues per turn in the alpha helix; in other words, the helix will repeat itself every 36 residues, with ten turns of the helix in that interval.

**Beta sheet.** The beta sheet involves H-bonding between backbone residues in adjacent chains. In the beta sheet, a single chain forms H-bonds with its neighboring chains, with the donor (amide) and acceptor (carbonyl) atoms pointing sideways rather than along the chain, as in the alpha helix. Beta sheets can be either parallel, where the chains point in the same direction when represented in the amino- to carboxyl- terminus, or antiparallel, where the amino- to carboxyl- directions of the adjacent chains point in the same direction. (See Figure 5-5.)

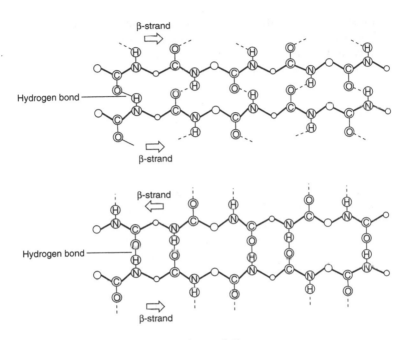

β-strand

Hydrogen bond

β-strand

β-strand

Hydrogen bond

β-strand

Figure 5-5

Different amino acids favor the formation of alpha helices, beta pleated sheets, or loops. The primary sequences and secondary structures are known for over 1,000 different proteins. Correlation of these sequences and structures revealed that some amino acids are found more often in alpha helices, beta sheets, or neither. Helix formers include alanine, cysteine, leucine, methionine, glutamic acid, glutamine, histidine, and lysine. Beta formers include valine, isoleucine, phenylalanine, tyrosine, tryptophan, and threonine. Serine, glycine, aspartic acid, asparagine, and proline are found most often in turns.

No relationship is apparent between the chemical nature of the amino acid side chain and the existence of amino acid in one structure or another. For example, Glu and Asp are closely related chemically (and can often be interchanged without affecting a protein's activity), yet the former is likely to be found in helices and the latter in turns.

Rationalizing the fact that Gly and Pro are found in turns is somewhat easier. Glycine has only a single hydrogen atom for its side chain. Because of this, a glycine peptide bond is more flexible than those of the other amino acids. This flexibility allows glycine to form turns between secondary structural elements. Conversely, proline, because it contains a secondary amino group, forms rigid peptide bonds that cannot be accommodated in either alpha or beta helices.

### Fibrous and globular proteins

The large-scale characteristics of proteins are consistent with their secondary structures. Proteins can be either fibrous (derived from fibers) or globular (meaning, like a globe). Fibrous proteins are usually important in forming biological structures. For example, collagen forms part of the matrix upon which cells are arranged in animal tissues. The fibrous protein keratin forms structures such as hair and fingernails. The structures of keratin illustrate the importance of secondary structure in giving proteins their overall properties.

Alpha keratin is found in sheep wool. The springy nature of wool is based on its composition of alpha helices that are coiled around and cross-linked to each other through cystine residues. Chemical reduction of the cystine in keratin to form cysteines breaks the cross-links. Subsequent oxidation of the cysteines allows new cross-links to form. This simple chemical reaction sequence is used in beauty shops and home permanent products to restructure the curl of human hair—the reducing agent accounts for the characteristic odor of these products. Beta keratin is found in bird feathers and human fingernails. The more brittle, flat structure of these body parts is determined by beta keratin being composed of beta sheets almost exclusively.

Globular proteins, such as most enzymes, usually consist of a combination of the two secondary structures—with important exceptions. For example, hemoglobin is almost entirely alpha-helical, and antibodies are composed almost entirely of beta structures. The secondary structures of proteins are often depicted in ribbon diagrams, where the helices and beta sheets of a protein are shown by corkscrews and arrows respectively, as shown in Figure 5-6.

**Hen egg-white lysozyme**

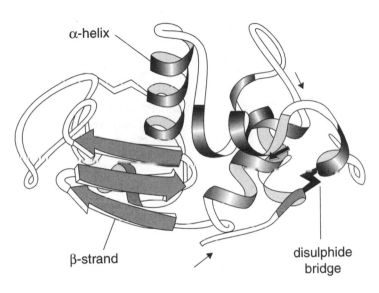

Figure 5-6

## Tertiary structure

The tertiary structure of a protein refers to the arrangement of amino acid side chains in the protein. *Generally, the information for protein structure is contained within the amino acid sequence of the protein itself.* This important principle of biochemistry was first determined by the biochemist Christian Anfinsen in studies of the enzyme ribonuclease. Ribonuclease catalyzes a simple hydrolysis of ribonucleic acid. The native enzyme has 124 amino acids; 8 of these are cysteines, forming 4 disulfide bonds. When ribonuclease was treated with mercaptoethanol to destroy the disulfide bonds and urea to disrupt its secondary and tertiary structure, all enzymatic activity was lost. Physical methods showed that this **denatured** form of ribonuclease had lost all detectable secondary and tertiary structure, although its amino acid sequence (primary structure) was intact. Anfinsen then slowly removed the urea and mercaptoethanol, then exposed the solution to air to reoxidize the cysteine pairs to disulfides. The renatured enzyme had full activity, leading to the conclusion that all the information required for the enzyme's three-dimensional structure was present only in the linear sequence of amino acids it contained and that the active structure of the enzyme was the thermodynamically most stable one.

This principle has been validated many times. For example, several enzymes, including ribonuclease, have been synthesized chemically from amino acids. The synthetic enzymes are fully active. Another validation has come from biotechnology. For example, human insulin, a hormone rather than an enzyme, can be made by yeast carrying the appropriate genes. Insulin made this way is indistinguishable from natural human insulin and is used extensively in treating diabetes.

**Protein tertiary structure.** Protein tertiary structures are the result of weak interactions. When a protein folds, either as it is being made on ribosomes or refolded after it is purified, the first step involves the formation of hydrogen bonds within the structure to nucleate secondary structural (alpha and beta) regions. For example, amide hydrogen atoms can form H-bonds with nearby carbonyl oxygens; an alpha helix or beta sheet can zip up, prompted by these small local structures.

**Hydrophobic interactions** among the amino acid side chains also determine tertiary structure. Most globular proteins have their hydrophobic side chains, for example, those of phenylalanine, valine, or tryptophan, located on the inside of the protein structure. Conversely, the hydrophilic amino acids, such as glutamic acid, serine, or asparagine, are generally found on the outside surface of the protein, where they are available for interaction with water. Alternatively, when these groups are found on the inside of soluble proteins, they often form **charge-charge interactions,** or **salt bridges,** bringing a positively charged side chain (such as Arg) close to a negative one (such as Glu).

In **membrane** proteins, these general principles are reversed: The hydrophobic amino acid side chains are found on the outside of the protein, where they are available to interact with the acyl groups of the membrane phospholipids, while the hydrophilic amino acids are on the inside of the protein, available for interacting with each other and with water-soluble species, such as inorganic ions. Membrane proteins are usually synthesized on membrane-bound ribosomes to facilitate their proper assembly and localization. Some membrane-bound proteins are synthesized on cytoplasmic ribosomes, with their hydrophobic residues inside the molecule, and they undergo a refolding when they later encounter the membrane where they will reside. **Van der Waal's** forces are important for a protein achieving its final shape. Although they are individually very weak, the sum of these interactions contributes substantial energy to the final three-dimensional shape of the protein.

### Protein-assisted folding

Proteins may assist the folding of other proteins. Although the native, active structure of a protein is thought to be the most stable one thermodynamically, it isn't always achieved in high yield when a protein is allowed to fold on its own. This can be true whether the protein is synthesized in vitro or in vivo—outside or inside a living body. Most cells contain a variety of proteins, called **chaperonins,** which facilitate the proper folding of newly synthesized or denatured proteins. Chaperonins use ATP energy to assist the refolding of proteins. Because proteins are often denatured by heat (think of a hard-boiled egg), many chaperonins are expressed at high levels during **heat shock** of cells. Fever is one physiological heat shock, and chaperonins are among the proteins that protect cellular proteins from denaturing during a fever.

Usually, disulfide bonds form after a protein has achieved a final tertiary structure. Because they are so strong, the premature formation of an incorrect hydrogen bond could force a protein into an inactive tertiary structure. For example, if the disulfides of ribonuclease are allowed to form when the protein is in a denatured state, less than 1% of the enzyme activity is recovered, indicating that only a small minority of the disulfides are correct.

In contrast, when the protein is allowed to form the proper tertiary structure before disulfide formation, essentially all the enzymatic activity is recovered. The **disulfide interchange** enzyme acts on newly made proteins, catalyzing the breakage and rejoining of disulfides in a protein. Combined with the action of the chaperonins, the enzyme helps the protein achieve its final, native state, with all the disulfides formed appropriately.

## Quaternary structures

Proteins associate with each other to form quaternary structures. Many proteins consist of more than one subunit. For example, hemoglobin has a molecular weight of 64,000 and is composed of four subunits, each of molecular weight 16,000. Two of the subunits are alike, and two are different. The enzyme tryptophan synthetase from *Escherichia coli,* which catalyzes the final two steps in the biosynthesis of that amino acid, consists of two nonidentical subunits, each of which catalyzes one reaction. Other enzymes contain regulatory and catalytic subunits. Still other enzymes consist of aggregates of two, three, or more identical subunits. The specific, noncovalent association of protein subunits is termed the quaternary structure of a protein. If the subunits are not identical, the association is called heterotypic. The association of identical subunits is termed homotypic.

The same forces that contribute to the structure of a single polypeptide also contribute to subunit interactions. Salt bridges, hydrogen bonds, hydrophobic, and van der Waal's interaction act in an additive fashion to specifically associate subunits.

# CHAPTER 6
## THE PHYSIOLOGICAL CHEMISTRY OF OXYGEN BINDING BY MYOGLOBIN AND HEMOGLOBIN

## The Chemistry of Molecular Oxygen

Metabolism can be either **aerobic** (requiring oxygen) or **anaerobic** (occurring in the absence of oxygen). Anaerobic metabolism is the older process: Earth's atmosphere has contained molecular oxygen for less than half the planet's existence. For organisms like yeast that can operate in either mode, aerobic metabolism is generally the more efficient process, yielding tenfold more energy from the metabolism of a molecule of glucose than do anaerobic processes. But the efficiency that comes from the use of molecular oxygen as an electron acceptor carries a price. Molecular oxygen is easily transformed into toxic compounds. For example, hydrogen peroxide, $H_2O_2$, is used as a disinfectant, as is ozone, $O_3$. Furthermore, molecular oxygen can also oxidize metal ions, and that can cause problems. Iron-containing enzymes and proteins use reduced iron, Fe(II) or Fe(I), and don't function if the iron atoms are oxidized to the stable Fe(III) form. Organisms must have means of preventing the oxidation of their iron atoms.

The third problem caused by the use of molecular oxygen as an electron acceptor is the fact that it really isn't very soluble in water. (If it were more soluble, people couldn't drown!) Multicellular organisms have evolved various oxygen transporters to solve the twofold problem of keeping oxygen tied up and less toxic as well as being able to deliver $O_2$ rapidly enough and in sufficient quantity to support metabolism. All animals (other than insects) with more than one kind of cell have evolved specialized proteins to carry oxygen to their tissues. The protein responsible for carrying oxygen in the blood of most terrestrial animals is **hemoglobin.** Within the tissues, especially muscle tissue, a related oxygen carrier, **myoglobin,** keeps molecular oxygen available for its final reduction to water as the end product of catabolism (nutrient utilization).

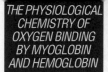
### Hemoglobin and myoglobin

Hemoglobin and myoglobin are only slightly related in primary sequence. (See Chapter 5 for more on sequences.) Although most amino acids are different between the two sequences, the amino acid changes between the two proteins are generally conservative. More strikingly, the secondary structures of myoglobin and the subunits of hemoglobin are virtually identical, as shown in Figure 6-1. Both proteins are largely alpha-helical, and the helices fit together in a similar way. One $O_2$ molecule is bound to each protein molecule by a coordinate covalent bond to an iron atom (Fe(II)) in the **heme group.** Heme is a square planar molecule containing four *pyrrole* groups, whose nitrogens form coordinate covalent bonds with four of the iron's six available positions. One position is used to form a coordinate covalent bond with the side chain of a single histidine amino acid of the protein, called the **proximal histidine.** The sixth and last orbital is used for oxygen. It is empty in the nonoxygenated forms of hemoglobin and myoglobin.

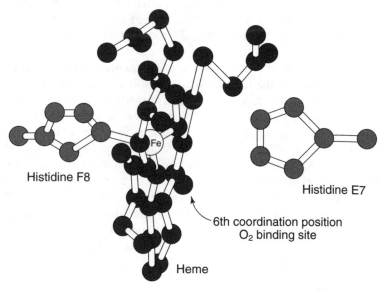

Figure 6-1

When molecular oxygen encounters an isolated heme molecule, it rapidly converts the Fe(II) to Fe(III). The oxidized heme binds oxygen very poorly. Obviously, if this happened to the Fe(II) groups of hemoglobin and myoglobin, the proteins would be less useful as oxygen carriers. Oxidation of the heme iron is prevented by the presence of the **distal histidine** side chain, which prevents the $O_2$ from forming a linear Fe–O–O bond. The bond between Fe and $O_2$ is bent, meaning that this bond is not as strong as it might be. *Weaker oxygen binding means easier oxygen release.* This is an important principle in understanding not only heme chemistry but also the regulation of hemoglobin's affinity for oxygen.

The differences between hemoglobin and myoglobin are most important at the level of quaternary structure. Hemoglobin is a tetramer composed of two each of two types of closely related subunits, alpha and beta. Myoglobin is a monomer (so it doesn't have a quaternary structure at all). Myoglobin binds oxygen more tightly than does hemoglobin. This difference in binding energy reflects the movement of oxygen from the bloodstream to the cells, from hemoglobin to myoglobin.

**Myoglobin binds oxygen**
The binding of $O_2$ to myoglobin is a simple equilibrium reaction:

$$Mb + O_2 \rightleftharpoons MbO_2$$

Each myoglobin molecule is capable of binding one oxygen, becausemyoglobin contains one heme per molecule. Even though the reaction of myoglobin and oxygen takes place in solution, it is convenient to measure the concentration of oxygen in terms of its **partial pressure,** the amount of gas in the atmosphere that is in equilibrium with the oxygen in solution.

The titration curve of myoglobin with oxygen is a hyperbola, as shown in Figure 6-2 of the form:

$$Y = pO_2 \, / \, pO_2 + P_{50}$$

where $Y$ is the fraction of oxygenated myoglobin, $pO_2$ is the partial pressure of $O_2$, expressed in torr (mm Hg; 760 torr = 1 atmosphere) and $P_0$ is the partial pressure of $O_2$ required to bind 50% of the myoglobin molecules. The derivation of this equation from the equilibrium constant for binding is not reproduced here but may be found in many standard chemistry and biochemistry textbooks. In the above equation, if $Y$ is set at 0.5, $P_{50} = pO_2$.

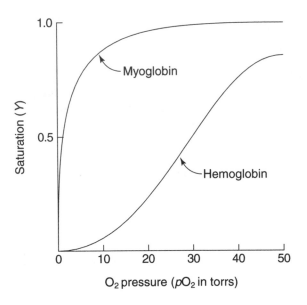

Figure 6-2

Under standard physiological conditions, the $P_{50}$ for $O_2$ binding to myoglobin is approximately 1 torr. Because the atmosphere is approximately 21% oxygen, at 1 atmosphere of pressure, $pO_2 = 0.21 \times 760 = 159$ torr myoglobin would be almost totally saturated:

$$Y = 159/(159 + 1) = 0.99$$

Because $pO_2$ in blood is lower than atmospheric $pO_2$, Y varies from 0.91 (venous blood) to 0.97 (arterial blood).

**Binding oxygen to hemoglobin**

Because hemoglobin has four subunits, its binding of oxygen can reflect multiple equilibria:

$$Hb + O_2 \ Hb{-}O_2$$

$$Hb{-}O_2 + s \ O_2 \ Hb{-}(O_2)_2$$

$$Hb{-}(O_2)_2 + O_2 \ Hb{-}(O_2)_3$$

$$Hb{-}(O_2)_3 + O_2 \ Hb{-}(O_2)_4$$

The equilibrium constants for these four $O_2$ binding events are dependent on each other and on the solution conditions. The influence of one oxygen's binding on the binding of another oxygen is called a **homotropic effect.** Overall, this **cooperative** equilibrium binding makes the binding curve **sigmoidal** rather than hyperbolic, as Figure 6-3 shows. The $P_{50}$ of hemoglobin in red blood cells is about 26 torr under normal physiological conditions. In the alveoli of the lungs, $pO_2$ is about 100 torr, and close to 20 torr in the tissues. So you may expect hemoglobin to be about 80% loaded in the lungs and a little over 40% loaded with oxygen in the tissue capillaries. In fact, hemoglobin can be more $O_2$-saturated in the lungs and less saturated in the capillaries, as the following sections explain.

Cooperativity is a complex subject; one model is the interconversion of the hemoglobin between two states—the T (tense) and the R (relaxed) conformations—of the molecule. The R state has higher affinity for oxygen. Under conditions where $pO_2$ is high (such as in the lungs), the R state is favored; in conditions where $pO_2$ is low (as in exercising muscle), the T state is favored.

Quantitatively, the binding curve of a complex protein like hemoglobin is described by the approximation:

$$Y = pO_2{}^n / pO2^n + P50^n$$

where $n$ is the number of interacting subunits. The equation can be manipulated into logarithmic form:

$$\log [Y/(1 - Y)] = n \log(pO_2)^+ - n\log P_{50}$$

A graph of this equation yields a **Hill plot** (named after the British physiologist who first described it), as shown in Figure 6-3. Hill plots of $\log [Y/(1 - Y)]$ versus $\log(pO_2)$ are linear, over at least part of the range, with slope $n$ (sometimes called a Hill coefficient), which is the minimal number of interacting subunits. For myoglobin, which only has one subunit, the slope must be 1; for hemoglobin, the Hill coefficient is 2.8, indicating a *minimum* of three interacting subunits, although there are four subunits in a molecule of hemoglobin.

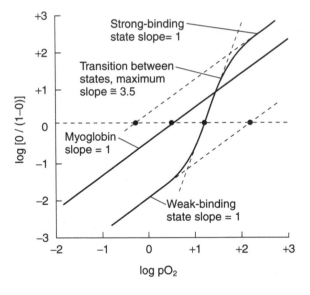

Figure 6-3

Even if no other factors operated, because the $P_{50}$ of hemoglobin is greater than the $P_{50}$ for myoglobin, *hemoglobin in the blood gives up $O_2$ to myoglobin in the tissues.* This follows from the principle discussed above, that a greater $P_{50}$ means lower affinity, and weaker binding means greater release.

**Physiological conditions and hemoglobin**

Anyone who goes from a long period of inactivity to vigorous exercise (for example, from months of watching television to several hours of racquetball) experiences stiffness due to the buildup of *lactic acid* in the tissues. Even during moderate exercise, muscle activity generates the weak acid carbon dioxide. For example, if glucose is oxidized to water and carbon dioxide and the enzyme **carbonic anhydrase** interconverts $CO_2$ and carbonic acid:

$$C_6H_{12}O_6 + 6\ O_2 \quad 6\ CO_2 + 6\ H_2O$$

$$CO_2 + H_2O \quad H_2CO_3 \quad H^+ + HCO_3^-$$

The net effect is a drop in pH due to metabolism.

**Acidic conditions and hemoglobin**

A decrease in pH *increases* the $P_{50}$ of hemoglobin. This phenomenon is called the **Bohr effect.** Because of the Bohr effect, more $O_2$ is released from hemoglobin to the tissues where it is needed than would be predicted from simple equilibrium effects. Conversely, in the lungs, where $CO_2$ leaves the bloodstream by diffusion, the pH increases relative to that in venous blood, and hemoglobin binds oxygen more tightly.

**Temperature**

Because heat is a product of metabolism, more oxygen needs to be delivered to the tissues when metabolism is very active, for example during vigorous exercise. Hemoglobin binds oxygen less tightly at higher temperature so that it gives up its oxygen more readily when it is needed.

**The regulatory compound, 2,3—bisphosphoglycerate (BPG) and hemoglobin**

BPG is a byproduct of glucose metabolism; its structure is shown in Figure 6-4. There is approximately one molecule of BPG per hemoglobin tetramer in the red blood cell. BPG is an **allosteric regulator;** it binds to a specific site on hemoglobin and shifts the dissociation curve to the left. This means that oxygen is delivered more readily to the tissues. BPG levels increase as an adaptation to high altitude (for example, on moving from Seattle at sea level to Denver at an altitude of 1,700 meters), allowing physical activity under low oxygen conditions. At still higher altitudes, where the $pO_2$ is lower still, BPG limits the ability of the hemoglobin to bind oxygen in the lungs. This may limit long-term human activity to altitudes below 5,000 meters above sea level— humans simply can't get enough oxygen into their hemoglobin if the $pO_2$ is lower than that found at that level.

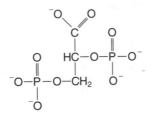

**Fetal hemoglobin**

In humans and other mammals, the developing embryo and fetus express different forms of hemoglobin than does the mother. The oxygen affinities of fetal hemoglobin are considerably greater than that of maternal hemoglobin. This phenomenon fits with the fact that fetal hemoglobin must be oxygenated in the placenta, where the $pO_2$ is lower than it is in the lungs. The higher affinity (lower $P_{50}$) of fetal hemoglobin is due to its lower affinity for BPG. Because BPG binding and $O_2$ binding interfere with each other, reduced affinity for the former means increased affinity for the latter. Fetal hemoglobin is replaced by the mature form in human infants by about six months of age.

# CHAPTER 7
## ENZYMES

## Enzymes Are Catalysts

A **catalyst** is a chemical that increases the rate of a chemical reaction without itself being changed by the reaction. The fact that they aren't changed by participating in a reaction distinguishes catalysts from **substrates,** which are the reactants on which catalysts work. Enzymes catalyze biochemical reactions. They are similar to other chemical catalysts in many ways:

1. Enzymes and chemical catalysts both affect the *rate but not the equilibrium constant* of a chemical reaction. Reactions proceed downhill energetically, in accord with the Second Law of Thermodynamics (see Chapter 3). Catalysts merely reduce the time that a thermodynamically favored reaction requires to reach equilibrium. Remember that the Second Law of Thermodynamics tells *whether* a reaction can occur but not how *fast* it occurs.

2. Enzymes and chemical catalysts increase the rate of a chemical reaction in *both directions,* forward and reverse. This principle of catalysis follows from the fact that catalysts can't change the equilibrium of a reaction. Because a reaction at equilibrium occurs at the same rate both directions, a catalyst that speeds up the forward but not the reverse reaction necessarily alters the equilibrium of the reaction.

3. Enzymes and chemical catalysts *bind their substrates,* not permanently, but transiently—for a brief time. There is no action at a distance involved. The portion of an enzyme that binds substrate and carries out the actual catalysis is termed the **active site.**

Enzymes differ from ordinary chemical catalysts in several important respects:

1. **Enzymes are specific.** Chemical catalysts can react with a variety of substrates. For example, hydroxide ions can catalyze the formation of double bonds and also the hydrolysis of esters. Usually enzymes catalyze only a single type of reaction, and often they work only on one or a few substrate compounds.

2. **Enzymes work under mild conditions.** Chemical catalysts often require high temperature and/or pressure to function. For example, nitrogen can be reduced to ammonia industrially by the Haber process, catalyzed by iron at 500° C. and at 300 atmospheres pressure of $N_2$. In contrast, the same reaction is carried out enzymatically at 25° C. and less than 1 atmosphere pressure of $N_2$. These gentle conditions of temperature, pressure, and pH characterize enzymatic catalysis, especially within cells.

3. **Enzymes are stereospecific.** Chemical catalysis of a reaction usually leads to a mixture of stereoisomers. For example, the addition of acid-catalyzed water to a double bond leads to an equimolar (50:50) mixture of D and L isomers where the water is added. In contrast, catalysis of water addition by enzymes results in complete formation of either the D or L isomer, but not both.

4. **Enzymes are macromolecules.** The macromolecules are composed of protein, or in a few cases, RNA. Most chemical catalysts are either surfaces, for example, metals like platinum, or else small ions, such as hydroxide ions.

5. **Enzymes are often regulated.** The regulation occurs either by the concentration of substrates, by binding small molecules or other proteins, or by covalent modification of the enzymes' amino acid side chains. Thus, an enzyme's effectiveness can

be altered without changing the concentration of the enzyme; on the other hand, the effectiveness of a chemical catalyst is generally determined by its overall concentration.

## Six Types of Enzyme Catalysts

Although a huge number of reactions occur in living systems, these reactions fall into only half a dozen types. The reactions are:

1. **Oxidation and reduction.** Enzymes that carry out these reactions are called **oxidoreductases.** For example, alcohol dehydrogenase converts primary alcohols to aldehydes.

   $H_3CCH_2OH + NAD$  $H_3CCHO + NADH + H^+$

   In this reaction, ethanol is converted to acetaldehyde, and the *cofactor,* NAD, is converted to NADH. In other words, ethanol is oxidized, and NAD is reduced. (The charges don't balance, because NAD has some other charged groups.) Remember that in redox reactions, one substrate is oxidized and one is reduced.

2. **Group transfer reactions.** These enzymes, called **transferases,** move functional groups from one molecule to another. For example, alanine aminotransferase shuffles the alpha-amino group between alanine and aspartate:

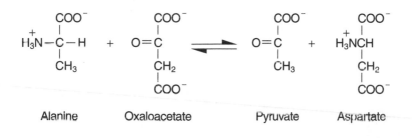

| Alanine | Oxaloacetate | Pyruvate | Aspartate |

Other transferases move phosphate groups between ATP and other compounds, sugar residues to form disaccharides, and so on.

3. **Hydrolysis.** These enzymes, termed **hydrolases,** break single bonds by adding the elements of water. For example, phosphatases break the oxygen-phosphorus bond of phosphate esters:

$$R-O-\overset{\overset{\displaystyle O}{\|}}{\underset{\underset{\displaystyle O^-}{|}}{P}}-O^- \quad \xrightarrow{\ H_2O\ } \quad R-OH$$

$$HO-\overset{\overset{\displaystyle O^+}{\|}}{\underset{\underset{\displaystyle O^-}{|}}{P}}-O^-$$

Other hydrolases function as digestive enzymes, for example, by breaking the peptide bonds in proteins.

4. **Formation or removal of a double bond with group transfer.** The functional groups transferred by these **lyase** enzymes include amino groups, water, and ammonia. For example, decarboxylases remove $CO_2$ from alpha- or beta-keto acids:

Dehydratases remove water, as in fumarase (fumarate hydratase):

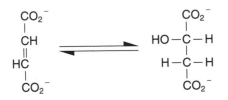

Deaminases remove ammonia, for example, in the removal of amino groups from amino acids:

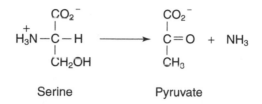

Serine    Pyruvate

5. **Isomerization of functional groups.** In many biochemical reactions, the position of a functional group is changed within a molecule, but the molecule itself contains the same number and kind of atoms that it did in the beginning. In other words, the substrate and product of the reaction are *isomers*. The **isomerases** (for example, triose phosphate isomerase, shown following), carry out these rearrangements.

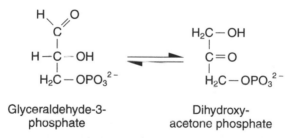

Glyceraldehyde-3-    Dihydroxy-
phosphate     acetone phosphate

6. **Single bond formation by eliminating the elements of water.** Hydrolases break bonds by adding the elements of water; **ligases** carry out the converse reaction, removing the elements of water from two functional groups to form a single bond. Synthetases are a subclass of ligases that use the hydrolysis of ATP to drive this formation. For example,

*aminoacyl-transfer RNA synthetases* join amino acids to their respective transfer RNAs in preparation for protein synthesis; the action of glycyl-tRNA synthetase is illustrated in this figure:

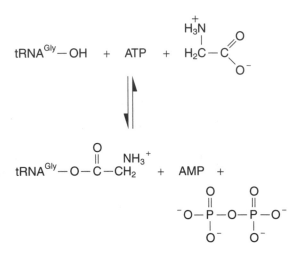

## The Michaelis-Menten equation

If an enzyme is added to a solution containing substrate, the substrate is converted to product, rapidly at first, and then more slowly, as the concentration of substrate decreases and the concentration of product increases. Plots of substrate (S) or product (P) against time, called **progress curves,** have the forms shown in Figure 7-1. Note that the two progress curves are simply inverses of each other. At the end of the reaction, equilibrium is reached, no net conversion of substrate to product occurs, and either curve approaches the horizontal.

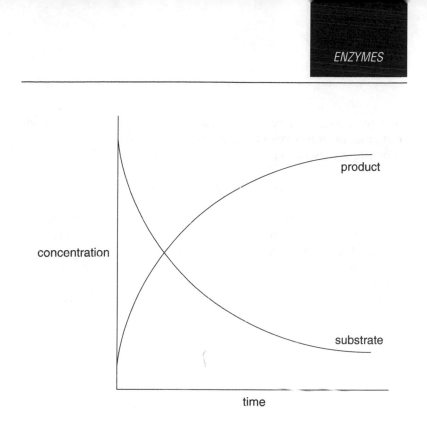

Figure 7-1

Another way to look at enzymes is with an **initial velocity** plot. The rate of reaction is determined early in the progress curve—very little product is present, but the enzyme has gone through a limited number of catalytic cycles. In other words, the enzyme is going through the sequence of product binding, chemical catalysis, and product release continually. This condition is called the **steady state**. For example, the three curves in Figure 7-2 represent progress curves for an enzyme under three different reaction conditions. In all three curves, the amount of enzyme is the same; however, the concentration of substrate is least in curve *(a)*, greater in curve *(b)*, and greatest in curve *(c)*. The progress curves show that more product forms as more substrate is added. The *slopes* of the progress curves at early time, that is, the rate of product formation with time also increase with increasing substrate concentration. These slopes, called the **initial rates** or **initial velocities,** of the reaction also increase as more substrate is present so that:

$$^v\text{curve } a \;{}^{<v}\text{curve } b \;{}^{<v}\text{curve } c$$

The more substrate is present, the greater the initial velocity, because enzymes act to bind to their substrates. Just as any other chemical reaction can be favored by increasing the concentration of a reactant, the formation of an enzyme-substrate complex can be favored by a higher concentration of substrate.

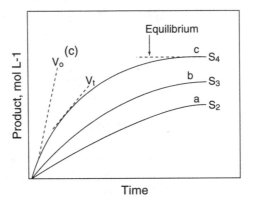

Figure 7-2

A plot of the initial velocities versus substrate concentration is a hyperbola (Figure 7-3). Why does the curve in Figure 7-3 flatten out? Because if the substrate concentration gets high enough, the enzyme spends all its time carrying out catalysis and no time waiting to bind substrate. In other words, the amount of substrate is high enough so that the enzyme is **saturated,** and the reaction rate has reached **maximal velocity, or $V_{max}$.** Note that the condition of maximal velocity in Figure 7-3 is *not the same as the state of thermodynamic equilibuium in Figures 7-1 and 7-2.*

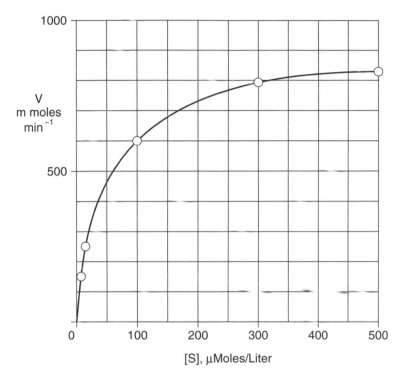

Figure 7-3

Although it is a velocity curve and not a binding curve, Figure 7-3 is a hyperbola—the same shape as the curve for the binding of $O_2$ to myoglobin in Chapter 6. Just as myoglobin is saturated with oxygen at high enough $pO_2$, so an enzyme is saturated with substrate at high enough substrate concentration, designated [S]. The equation describing the plot in Figure 7-2 is similar in form to the equation used for $O_2$ binding to myoglobin:

$$v = V_{max} \times [S]/K_m + [S]$$

$K_m$ is the **Michaelis constant** for the enzyme binding substrate. The Michaelis constant is analogous to, but *not* identical to, the binding constant for the substrate to the enzyme. $V_{max}$ is the **maximal velocity** available from the amount of enzyme in the reaction mixture. If you add more enzyme to a given amount of substrate, the velocity of the reaction (measured in moles of substrate converted per time) increases, because the increased amount of enzyme uses more substrate. This is accounted for by the realization that $V_{max}$ depends on the total amount of enzyme in the reaction mixture:

$$V_{max} = k_{cat} \times E_t$$

where $E_t$ is the total concentration of the enzyme and $k_{cat}$ is the rate constant for the slowest step in the reaction.

Other concepts follow from the Michaelis-Menten equation. When the velocity of an enzymatic reaction is one-half the maximal velocity:

$$v = V_{max}/2$$

then:

$$[S] = K_m$$

because:

$$v = V_{max} \times [S]/[S] + [S] = V_{max} \times [S]/2[S] = V_{max}/2$$

In other words, the $K_m$ is numerically equal to the amount of substrate required so that the velocity of the reaction is half of the maximal velocity.

Alternatively, when the concentration of substrate in the reaction is very high ($V_{max}$ conditions), then $[S] \gg K_m$, and the $K_m$ term in the denominator can be ignored in the equation, giving:

$$v = V_{max} \times [S]/[S] = V_{max}$$

On the other hand, when [S] $\ll$ $K_m$, the term [S] in the denominator of the Michaelis-Menten equation can be ignored, and the equation reduces to:

$$v = V_{max} \times [S]/K_m$$

In the last case, the enzyme is said to be under *first order* conditions, because the velocity depends directly on the concentration of substrate.

## Inhibitors of enzyme-catalyzed reactions

In the terms of the Michaelis-Menten equation, inhibitors can raise $K_m$, lower $V_{max}$, or both. Inhibitors form the basis of many drugs used in medicine. For example, therapy for high blood pressure often includes an inhibitor of the angiotensin converting enzyme, or ACE. This enzyme cleaves (hydrolyzes) angiotensin I to make angiotensin II. Angiotensin II raises blood pressure, so ACE inhibitors are used to treat high blood pressure. Another case is acetylsalicylic acid, or aspirin. Aspirin successfully treats inflammation because it covalently modifies, and therefore inactivates, a protein needed to make the signaling molecule that causes inflammation.

The principles behind enzyme inhibition are illustrated in the following examples.

Alkaline phosphatase catalyzes a simple hydrolysis reaction:

$$R-O-PO_3^{2-} + H_2O \rightarrow HPO_4^{2-} + ROH$$

Phosphate ion, a product of the reaction, also inhibits it by binding to the same phosphate site used for binding substrate. When phosphate is bound, the enzyme cannot bind substrate, so it is *inhibited* by the phosphate. How to overcome the inhibitor? Add more substrate: $R-O-PO_3^{2-}$. Because the substrate and the inhibitor bind to the same site on the enzyme, the more substrate that binds, the less inhibitor binds. When is the most substrate bound to the enzyme? Under $V_{max}$ conditions. Phosphate ion reduces the velocity of the alkaline phosphate reaction without reducing $V_{max}$. If velocity decreases, but $V_{max}$ doesn't, the only other thing that can change is $K_m$. Remember that $K_m$ is the concentration where $v = V_{max}/2$. Because more substrate is required to achieve $V_{max}$, $K_m$ must necessarily increase. This type of inhibition, where $K_m$ increases but $V_{max}$ is unchanged, is called **competitive** because the inhibitor and substrate compete for the same site on the enzyme (the active site).

Other cases of inhibition involve the binding of the inhibitor to a site other than the site where substrate binds. For example, the inhibitor can bind to the enzyme on the outside of the protein and thereby alter the tertiary structure of the enzyme so that its substrate binding site is unable to function. Because some of the enzyme is made nonfunctional, adding more substrate can't reverse the inhibition. $V_{max}$, the kinetic parameter that includes the $E_t$ term, is reduced. The binding of the inhibitor can also affect $K_m$ if the enzyme-inhibitor complex is partially active. Inhibitors that alter both $V_{max}$ and $K_m$ are called **noncompetitive;** the rare inhibitors that alter $V_{max}$ only are termed **uncompetitive.**

You can visualize the effects of inhibitors using reciprocal plots. If the Michaelis-Menten equation is inverted:

$$v = V_{max} \times [S]/K_m + [S]$$

$$1/v = (K_m + [S])/V_{max} \times [S] = (K_m/V_{max})(1/[S]) + 1/V_{max}$$

This equation is linear and has the same form as:

$$y = ax + b$$

so that a plot of $1/v$ versus $1/[S]$ (a **Lineweaver-Burk plot,** shown in Figure 7-4) has a slope equal to $K_m/V_{max}$ and a y-intercept equal to $1/V_{max}$. The x-intercept of a Lineweaver-Burk plot is equal to $^-1/K_m$.

**Competitive inhibitors** decrease the velocity of an enzymatic reaction by increasing the amount of substrate required to saturate the enzyme; therefore, they increase the apparent $K_m$ but do not affect $V_{max}$. A Lineweaver-Burk plot of a competitively inhibited enzyme reaction has an increased slope, but its intercept is unchanged.

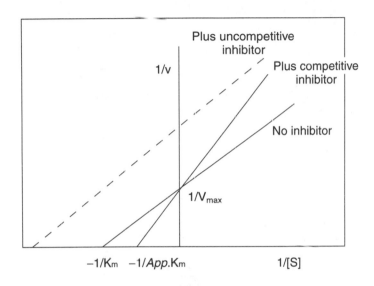

Figure 7-4

**Noncompetitive inhibitors** both increase the apparent $K_m$ and reduce the apparent $V_{max}$ of an enzyme-catalyzed reaction. Therefore, they affect both the slope and the y-intercept of a Lineweaver-Burk plot, as Figures 7-5 and 7-6 show. Uncompetitive inhibitors, because they reduce $V_{max}$ only, increase the reciprocal of $V_{max}$. The lines of the reciprocal plot are parallel in this case.

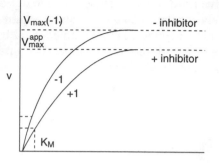

Figure 7-5

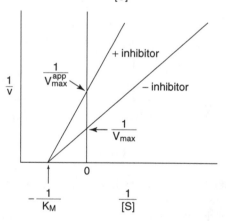

Figure 7-6

**Covalent inhibition** involves the chemical modification of the enzyme so that it is no longer active. For example, the compound diisopropylfluorophosphate reacts with many enzymes by adding a phosphate group to an essential serine hydroxyl group in the enzymes' active sites. When phosphorylated, the enzyme is totally inactive. Many useful pharmaceutical compounds work by covalent modification. Aspirin is a covalent modifier of enzymes involved in the inflammatory response. Penicillin covalently modifies enzymes required for bacterial cell-wall synthesis, rendering them inactive. Because the cell wall is not able to protect the bacterial cell, the organism bursts easily and is killed.

## Chemical Mechanisms of Enzyme Catalysis

How does an enzyme accomplish its tremendous enhancement of a reaction's rate (as much as a billion-fold)? There is an upper limit to the activity of an enzyme: It cannot operate faster than the rate at which it encounters the substrate. In solution, this rate is approximately $10^8$ to $10^9$ times per second ($sec^{-1}$). In the cell, enzymes acting on similar pathways are often located next to one another so that substrates don't have to diffuse away from one enzyme to the next—a mechanism that allows the enzymes to be more efficient than the theoretical limit. Even in solution, however, enzymes are powerful catalysts, and a variety of mechanisms bring about that power.

### The transition state

When a chemical reaction occurs, the energy content of the reacting molecule or atom increases. This is why most chemical reactions, whether they release heat or absorb heat, happen faster as the temperature is raised. The high-energy state of the reactants is called the **transition state.** For example, in a bond-breaking reaction, the transition state may be one where the reacting bond, although not completely broken, is vibrating at a frequency high enough that it is equally likely to split apart as to reform. Forming reactants or products results in the loss of energy from the transition state. This principle is shown in Figure 7-7, where the increased energy of the transition state is represented as a hill or barrier on the energy diagram. *Catalysts reduce the height of the barrier for achieving the transition state.*

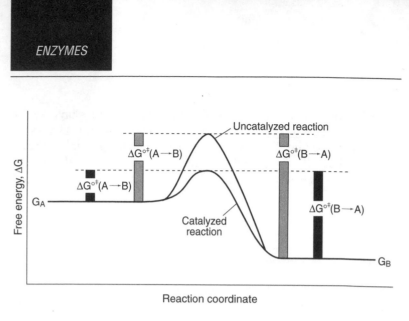

Figure 7-7

What are the chemical mechanisms that enzymes use to make it easier to get to the transition state? Enzymologists have determined that a number of mechanisms seem to operate, including:

1. **Proximity.** Enzymes can bring two molecules together in solution. For example, if a phosphate group is to be transferred from ATP to glucose, the probability of the two molecules coming close together is very low in free solution. After all, there are many other molecules that the ATP and the sugar could collide with. If the ATP and the sugar can bind separately and tightly to a third component—the enzyme's **active site**—the two components can react with each other more efficiently.

2. **Orientation.** Even when two molecules collide with enough energy to cause a reaction, they don't necessarily form products. They have to be oriented so that the energy of the colliding molecules is transferred to the reactive bond. Enzymes bind substrates so that the reactive groups are steered to the direction that can lead to a reaction.

3. **Induced fit.** Enzymes are flexible. In this regard, they are different from solid catalysts, like the metal catalysts used in chemical hydrogenation. After an enzyme binds its substrate(s), it changes conformation and forces the substrates into a strained or distorted structure that resembles the transition state. For example, the enzyme hexokinase closes like a clamshell when it binds glucose. In this conformation, the substrates are forced into a reactive state.

4. **Reactive amino acid groups.** The side chains of amino acids contain a variety of reactive residues. For example, histidine can accept and/or donate a proton to or from a substrate. In hydrolysis reactions, an acyl group can be bound to a serine side chain before it reacts with water. Having enzymes with these catalytic functions close to a substrate increases the rate of the reactions that use them. For example, a proton bound to histidine can be donated directly to a basic group on a substrate.

5. **Coenzymes and metal ions.** Besides their amino acid side chains, enzymes can provide other reactive groups. Coenyzmes are biomolecules that provide chemical groups that help catalysis. Like enzymes themselves, coenzymes are not changed during catalysis. This distinguishes them from other substrates, such as ATP, which are changed by enzyme action. Coenzymes, however, are not made of protein, as are most enzymes. Metal ions can also be found in the active sites of a number of enzymes, bound to the enzyme and sometimes to the substrate.

Coenzymes provide chemical functional groups that proteins lack. For example, only sulfhydryl groups on amino acids are able to participate in oxidation and reduction reactions, and the formation/breakage of disulfides does not provide enough reducing power to alter most biomolecules's functional groups. Electron transfer requires one of several coenzymes, usually either nicotinamide adenine dinucleotide, NAD, or flavin adenine dinucleotide, FAD, as electron acceptors and donors. Table 7-1 shows some of these coenzymes.

## Table 7-1 Selected Coenzymes

| *Coenzyme* | *Major Function* |
|---|---|
| Nicotinamide adenine dinucleotide (NAD$^+$) | Oxidation-reduction reactions |

*Structure*

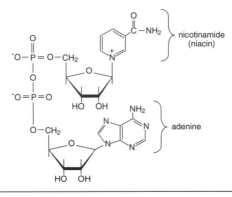

| *Coenzyme* | *Major Function* |
|---|---|
| Flavin adenine dinucleotide (FAD) | Oxidation-reduction reactions |

*Structure*

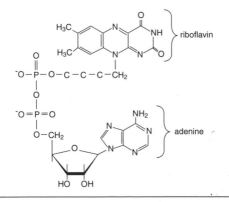

| *Coenzyme* | *Major Function* |
|---|---|
| Pyridoxal phosphate | Group-transfer reactions |

*Structure*

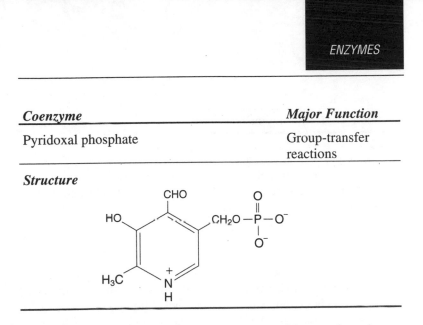

Some coenzymes participate in group-transfer reactions that are difficult to carry out with amino acid side chain chemistries alone. For example, none of the side chains of the normal 20 amino acids can accept an amino group easily. On the other hand, the coenzyme pyridoxal phosphate has a carbonyl group that is well adapted to accepting or donating amino groups.

**Vitamin conversion**
Vitamins are organic compounds required for human and animal growth. Many microorganisms (although by no means all) can grow and reproduce in a simple medium of sugars and inorganic salts. Likewise, photosynthetic organisms can synthesize all the organic molecules needed for life. These organisms don't need vitamins because they can synthesize them from simpler chemicals.

Our species has lost the ability to make vitamins. Thus, deficiency of niacin (nicotinamide), the "N" in NAD, leads to the disease *pellagra,* a collection of skin, intestinal, and neurological symptoms. (Niacin can be synthesized from the amino acid tryptophan, so pellagra results from a deficiency of both niacin and tryptophan in the diet.)

Chymotrypsin: An Enzyme at Work

The principles of enzyme action are illustrated by the enzyme chymotrypsin. Chymotrypsin digests proteins in the intestine by hydrolyzing the peptide bond at the carboxy side (to the right as conventionally written) of a hydrophobic amino acid. Thus, the small peptide glycylphenylalanylglycine (GlyPheGly) is hydrolyzed to GlyPhe and Gly.

The active site of chymotrypsin contains several reactive groups in close proximity to the binding site for the hydrophobic amino acid side chain. This binding site is a deep pocket lined with hydrophobic amino acid side chains. A charged amino acid side chain has to give up its favorable interactions with water to insert into the binding pocket, but the hydrophobic side chain of phenylalanine, for example, gains a favorable interaction and leaves water behind. When the substrate is bound to the enzyme, nearby amino acid side chains of the active site participate in the enzymatic reaction. Figure 7-8 shows the process inside such a pocket.

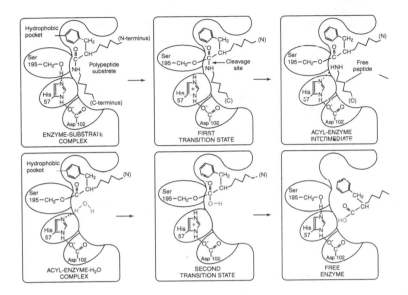

Figure 7-8

Three amino acid side chains participate in the catalytic reaction by forming a **charge relay system.** One amino acid, serine, is partially deprotonated by a nearby histidine. Normally, histidine is not a strong enough base to remove a proton from serine—but the histidine is itself partially deprotonated by the carboxylate side chain of an aspartate. The end result of this charge relay system is that serine is able to attack the carbonyl carbon, breaking the peptide bond and forming an **acyl enzyme intermediate.** The proton originally bound to the serine hydroxyl group is transferred to the amino group in the peptide bond, leaving histidine able to accept a proton from water. The rest of the water attacks the acyl enzyme intermediate, leading to the reforming of the original enzyme.

The importance of these amino acid side chains is illustrated by the action of two kinds of irreversible enzyme inhibitors (shown in Figure 7-9). Diisopropylfluorophosphate transfers its phosphate to the active site serine. The resulting phospho-enzyme is totally inactive. Chloromethyl ketones alkylate the active site histidine.

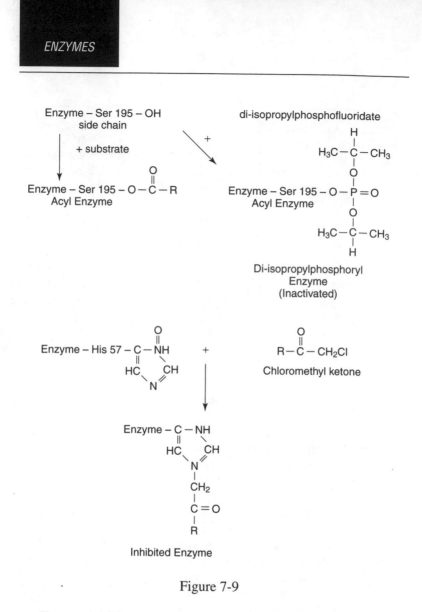

Figure 7-9

Enzyme inhibitors are often poisonous. For example, diisopropyl-fluorophosphate is a nerve poison because the enzyme acetylcholinesterase has a reactive site serine. Chymotrypsin and acetylcholinesterase are both members of the class of enzymes known as **serine esterases,** which are all inhibited by diisopropylfluorophosphate.

## Enzyme Regulation

Many of the biochemical reactions in a living cell can go both ways. For example, mammalian cells both catabolize and synthesize glucose. The rates at which these reactions occur must be regulated; otherwise, energy is wasted by what is called a **futile cycle** carrying out opposing reactions at high rates with no net substrate flow in either direction. Remember that the Second Law of Thermodynamics states that entropy increases in a favored reaction; entropy is wasted energy in that it can't be used to carry out work. Sometimes an enzyme that uses ATP as a substrate to transfer phosphate to another molecule can hydrolyze ATP to ADP and inorganic phosphate in the absence of the other substrate. This kind of reaction would obviously consume the cell's energy without doing useful work.

### Allostery and enzyme regulation

**Allostery** is the change in the kinetic properties of an enzyme caused by binding to another molecule. The binding of a small molecule to the enzyme alters its conformation so that it carries out catalysis more or less efficiently. For example, the binding of one molecule of a substrate to an enzyme can cause it to undergo a conformational change so that it binds the next molecule of substrate more efficiently. The first conformation is termed the **T (tense) state;** the second is called the **R (relaxed) state.** In other words, higher concentrations of substrate favor the *conversion of the T state to the R state.* This special case of regulation by substrate concentration is called **cooperativity.** Another case of cooperativity that has already been discussed in Chapter 6 is the cooperative binding of oxygen to hemoglobin. Cooperativity only operates on enzymes with more than one subunit.

Similarly, enzymes can be allosterically regulated by association with other molecules. Often the first enzyme in a metabolic pathway is **feedback-inhibited** by the product of that pathway. For example, anthranilate synthetase, the first enzyme in the biosynthesis of tryptophan, is inhibited by tryptophan, but not by other amino acids.

Other small molecules can act as **feed-forward activators.** For example, in DNA synthesis, the amounts of purine and pyrimidine nucleotides must be kept roughly equal. The enzyme aspartate transcarbamoylase is feedback-inhibited by CTP, the product of its metabolic pathway, and feed-forward-activated by ATP. If there is an excess of pyrimidines, inhibition by CTP slows the reaction, while if there is an excess of purines, ATP activates the enzyme, ultimately increasing the amount of pyrimidines in the cell.

This concept is similar to the conformational changes that occur in hemoglobin in response to changes in pH, fructose bisphosphate (FBP), and other conditions. Kinetically, these effects can be described by Hill plots where $\log(v/V_{max})/(1 - v/V_{max})$ is plotted versus $\log[S]$. (See Chapter 6.) The magnitude of the slope gives the minimal number of independent binding sites.

## Covalent Modification

Enzymes can be regulated by transfer of a molecule or atom from a donor to an amino acid side chain that serves as the acceptor of the transferred molecule. Another way of regulating an enzyme is by altering the amino acid sequence itself by proteolytic cleavage.

### Phosphorylation

Many enzymes are activated or inactivated by the transfer of inorganic phosphate from ATP to an acceptor—for example, the side-chain oxygen of serine. The combination of protein phosphorylation by kinases and dephosphorylyation by phosphatases can afford a fine level of control over enzyme activity.

## Zymogens

Why doesn't chymotrypsin digest us from the inside out? After all, it can digest many proteins, including our own. One answer is that chymotyrpsin is synthesized in the pancreas as an inactive form, termed **chymotrypsinogen,** in which form it is transported to the duodenum (the part of the small intestine, located just after the stomach). These inactive forms of enzymes are called **zymogens.** Chymotrypsinogen in the duodenum is converted to chymotrypsin by the related serine protease trypsin, which prefers to cut proteins at the carboxy side of basic amino acids. Two small segments of chymotrypsinogen are removed to make chymotrypsin.

Many proteins are synthesized in an inactive form and are only activated at a particular site in the cell or the body. Besides digestive enzymes, the proteins (called factors) involved in blood clotting circulate in the bloodstream as zymogens. The first proteins in the pathways are activated by trauma or tissue damage. The activated proteins cleave other zymogens to active enzymes, which in turn cleave other factors. Ultimately, the protein fibrinogen is cleaved to fibrin, the protein component of the clot.

There are two types of metabolic pathways: **catabolic,** involving the breakdown of biochemicals into simpler compounds, and **anabolic,** involving the synthesis of biochemicals from simpler molecules. Each living cell has thousands of distinct metabolic reactions. Each reaction is catalyzed by an enzyme and is linked to other reactions through a pathway. How can you keep them all straight? It is nearly impossible to memorize them. The purpose of this chapter is to provide an organizational framework to metabolism that allows you to view it as something other than a collection of disjointed pathways.

## Metabolism: A Collection of Linked Oxidation and Reduction Processes

Photosynthetic organisms *fix* $CO_2$ to form organic molecules, such as glucose. The carbon in $CO_2$ is in the +4 oxidation state, while the carbon in glucose ($C_6H_{12}O_6$) has an oxidation state of zero. Thus, carbon fixation must involve the reduction of carbon. Along with this process, something else must be oxidized—the oxygen of water is converted to molecular oxygen. In chemical terms, the oxygen in water is oxidized from the -2 state to zero—the oxidation state of elemental $O_2$. If the glucose is used by muscle cells during exercise, it can be broken down **aerobically** (with the participation of molecular oxygen) to $CO_2$ and water (effectively the reverse of photosynthesis), or **anaerobically** (without molecular oxygen being involved) to lactic acid, both of which are represented in Figure 8-1.

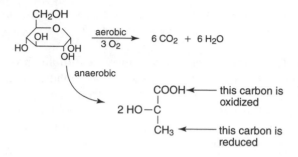

Figure 8-1

Lactic acid's empirical formula is $C_3H_6O_3$, so the carbons have a net oxidation number of zero, the same as in glucose ($C_6H_{12}O_6$). However, the carbons in lactic acid do not have the same oxidation number. The carboxyl carbon of lactic acid is more oxidized (+3) than the methyl carbon at the other end, which has three of its four bonds to hydrogen and therefore has an oxidation number of -3. A key feature of metabolic pathways is that *the oxidation of one component is balanced by the reduction of another.* The net result is that no electrons are lost or gained in the process.

**Energy production**

The primary energy-releasing pathways of metabolism involve the breakdown and synthesis of carbohydrates, lipids, and amino acids. The central metabolic pathway of the cell is the **tricarboxylic acid (TCA) cycle,** sometimes called the **Krebs cycle** after the German/British biochemist who recognized that the reactions of a limited number of dicarboxylic and tricarboxylic acids involved intermediates that were regenerated in each reaction series. Other pathways were known before that time, but they all seemed to have a definite beginning (substrate) and end (product). In contrast, 2-carbon compounds enter the Krebs cycle by reacting with a 4-carbon dicarboxylic acid to make a 6-carbon tricarboxylic acid, citric acid. (Another name for the tricarboxylic acid cycle is the **citric acid cycle.** This term refers

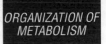
to the most important step for the entry of carbon-containing compounds into the cycle.) Citric acid is then oxidized in two successive steps, removing one carbon at a time, to make a four-carbon dicarboxylic acid, which then is metabolized to the first acceptor. The carbon count is therefore:

$$4 + 2 = 6; 6 = 5 + 1; 5 = 4 + 1$$

This shorthand describes the synthesis of a six-carbon compound from four- and two-carbon substrates and the subsequent breakdown of the six-carbon compound. This numerical shorthand will be used throughout this series.

The breakdown of citric acid involves the oxidation of carbon to $CO_2$ and the reduction of a cofactor, NAD, which then sends its electrons through a series of reactions to a **terminal electron acceptor.** In terrestrial animals and plants, this acceptor is oxygen, which forms water. Other organisms, especially bacteria, can use a variety of terminal electron acceptors—for example, sulfur or organic compounds.

### Catabolic pathways feed into the TCA cycle
Breakdown of carbohydrates and lipids leads to the synthesis of two-carbon intermediates of the Krebs cycle. Breakdown of amino acids leads to the synthesis of either two- or four-carbon compounds that can enter the Krebs cycle. Purines and pyrimidines are generally not broken down but rather are recycled by animals and plants, although purines and pyrimidines can be broken down into carbon dioxide and ammonia by plants and bacteria.

### Biosynthetic reactions versus catabolic reactions
The synthesis of large molecules from smaller compounds involves the net reduction of carbon, as in the synthesis of glucose from carbon dioxide during photosynthesis, or the synthesis of lipids from carbohydrate by animals. If synthesis and breakdown were the exact opposites of each other, there would be no way for an organism to

carry out net synthesis or degradation. This problem is a consequence of the laws of thermodynamics, which state that the free energy change of a given reaction is constant and that favored reactions are characterized by products having lower free energy than reactants. Therefore, if a reaction has a negative free energy in one direction, it will have a positive free energy, that is, be unfavored, in the other direction.

Organisms get around this problem by using different reactions in one or more steps of the overall pathway, depending on whether the reaction is proceeding in a catabolic or anabolic direction. The mammalian liver can both break down glucose for energy and synthesize glucose from food (amino acids, for example). Different reactions and enzymes participate in the catabolic (breakdown) and anabolic (synthetic) directions. The anabolic direction involves the input of energy in the form of ATP hydrolysis. This **coupling** of ATP hydrolysis to an otherwise unfavored reaction is a general theme in biochemistry.

Probably the oldest biochemical reaction known to humans is brewing. It may even predate agriculture—the earliest humans didn't have to cultivate grapes to notice that certain fruits were intoxicating after a period of fermentation. The remains of beers and recipes for brewing have been found in the Egyptian pyramids and in Mesopotamia. Brewing predates the second-oldest reaction, soap-making, by centuries: Draw your own conclusions!

Therefore, it's somewhat fitting that biochemistry began as a science 100 years ago with the demonstration by the Büchner brothers that sucrose could be fermented into ethanol using a yeast-cell extract. The study of glycolysis led to many of the concepts discussed in this book, including the roles of ATP, cofactors, and enzyme regulation.

**Glycolysis** describes the breakdown of a 6-carbon carbohydrate to two molecules of a 3-compound carboxylic acid:

$$6 \rightarrow 2(3)$$

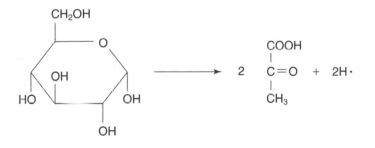

The 3-carbon acid, pyruvic acid, is then further metabolized, either aerobically or anaerobically:

**Aerobic:** $3 \rightarrow 2 \sim + 1$

$$\begin{array}{l} \text{COOH} \\ | \\ \text{C}=\text{O} \\ | \\ \text{CH}_3 \end{array} \longrightarrow \quad CO_2 \quad + \quad H_3C - \overset{\overset{\displaystyle O}{\|}}{C} \sim \text{Coenzyme A}$$

The $\sim$ refers to the binding of the 2-carbon compound with a special coenzyme, Coenzyme A. The 1-carbon product is carbon dioxide. Pyruvate can also be metabolized anaerobically, in which case it receives electrons that were initially removed during glycolysis:

**Animals (anaerobic):** $3 + 2e^- \rightarrow 3$

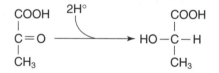

**Yeast (anaerobic):** $3 \rightarrow 2 + 1; 2 + 2\,e- \rightarrow 2$

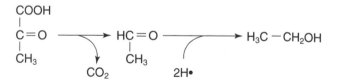

In animals, lactic acid is transported out of the muscles to the liver, where it is reconverted into glucose or amino acids. Lactic acid buildup is a source of the muscle stiffness that occurs after vigorous exercise. In yeast, pyruvate is converted to acetaldehyde and $CO_2$; then the acetaldehyde is reduced to ethanol.

Glycolysis can be divided into two parts, depending on whether the reactions consume or generate ATP. Enzyme nomenclature for the glycolytic pathway can be confusing. The names are historical rather than systematic and usually reflect the way the enzyme can be assayed. For example, the reduction of pyruvate to lactate is catalyzed by the enzyme lactate dehydrogenase, even though the relevant physiological reaction in glycolysis is the reduction of pyruvate, not the oxidation of lactate. (See the brief description of anaerobic metabolism in animals, in the preceding section.)

## Six-Carbon Reactions

Six-carbon reactions of glycolysis represent an energy investment of two high-energy phosphate bonds. Glucose enters glycolysis in a phosphorylated form, as glucose-6-phosphate:

When the glucose originates by breakdown of its polymeric forms, starch or glycogen, it is already phosphorylated, as glucose-1-phosphate, and the initial reaction is catalyzed by the enzyme **phosphoglucomutase.**

When glucose is present in its unphosphorylated form, the first reaction of glycolysis is a phosphorylation. Although the goal of glycolysis is the synthesis of ATP, a high-energy phosphate must be invested first, catalyzed by the enzyme **hexokinase:**

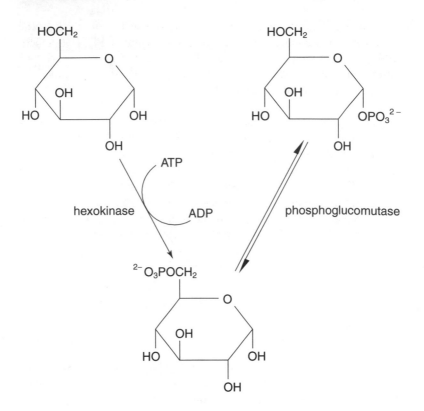

The end result of either of these reaction schemes is glucose-6-phosphate, which is now isomerized to fructose-6-phosphate by the enzyme **phosphoglucose isomerase:**

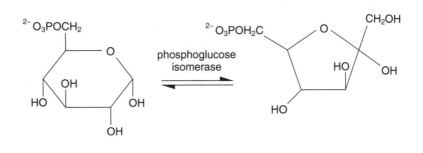

This is a reversible reaction, both *in vitro* and *in vivo*. Fructose-6-phosphate is then phosphorylated by the enzyme **phosphofructokinase,** in a second energy investment involving ATP:

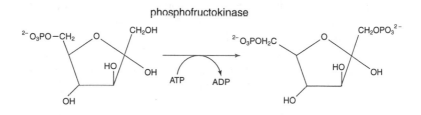

Fructose-1,6-bisphosphate, the product of this reaction, is then cleaved by **aldolase** to two 3-carbon compounds:

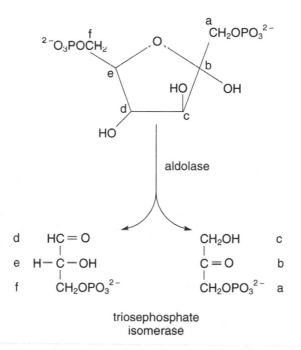

aldolase

| d | $HC=O$ | | $CH_2OH$ | c |
| e | $H-C-OH$ | | $C=O$ | b |
| f | $CH_2OPO_3{}^{2-}$ | | $CH_2OPO_3{}^{2-}$ | a |

triosephosphate
isomerase

These two products, glyceraldehyde-3-phosphate and dihydroxyacetone phosphate, are rapidly interconverted by **triose phosphate isomerase.**

The $K_{eq}$ of the aldolase reaction favors dihydroxyacetone phosphate; however, glyceraldehyde-3-phosphate is drained off for the further reactions of glycolysis, while dihydroxyacetone phosphate is not. This means that the concentration of glyceraldehyde-3-phosphate is very low during active metabolism and dihydroxyacetone phosphate is converted into glyceraldehyde-3-phospate.

The production of the triose phosphates represents the end of the investment. The reactions so far can be summarized as:

Glucose + 2 ATP → 2 triose phosphates + 2 ADP

In shorthand:

6 + 2 ATP → 3-P + 3-P + 2 ADP

---

Glycolysis, ATP, and NADH

The energy-yielding steps of glycolysis involve reactions of 3-carbon compounds to yield ATP and reducing equivalents as NADH. The first substrate for energy production is glyceraldehyde-3-phosphate, which reacts with ADP, inorganic phosphate, and NAD in a reaction catalyzed by the enzyme **glyceraldehyde-3-phosphate dehydrogenase:**

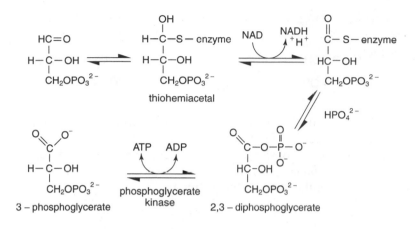

The reaction has several steps. In the first, a thiol carbon of the enzyme attacks the aldehyde carbon of glyceraldehyde-3-phosphate to make a thiohemiacetal intermediate. (Recall from organic chemistry that carbonyl carbons are electron-poor and therefore can bond with nucleophiles, including thiols from which the proton is removed.) Next, NAD accepts two electrons from the enzyme-bound glyceraldehyde-3-phosphate. The aldehyde of the substrate is *oxidized* to the level of a carboxylic acid in this step. Inorganic phosphate then displaces the thiol group at the oxidized carbon (carbon 1 of glyceraldehyde-3-phosphate) to form 1,3-bisphosphoglycerate:

The next step is the transfer of phosphate from 1,3-bisphosphoglycerate to ADP, making ATP, catalyzed by **phosphoglycerate kinase.**

This phase of glycolysis brings the energy balance from glucose back to zero. Two ATP phosphates were invested in making fructose-1,6-bisphosphate and two are now returned, one from each of the 3-carbon units resulting from the aldolase reaction.

The next reaction is the isomerization of 3-phosphoglycerate to 2-phosphoglycerate, catalyzed by **phosphoglycerate mutase:**

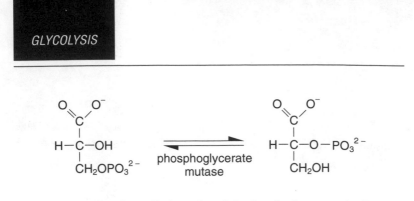

The reaction is pulled to the right by further metabolism of 2-phosphoglycerate. First, the compound is dehydrated by the removal of the hydroxyl group on carbon 3 and a proton from carbon 2, leaving a double bond between carbons 2 and 3. The enzyme responsible for this step is a lyase, **enolase:**

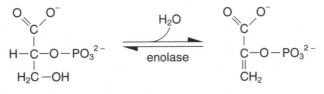

3 – phosphoglycerate                    phosphoenolpyruvate

Glyceraldehyde-3-phosphate + NAD + $P_i$ × 3-phosphoglyceric acid + NADH + $H^+$

Enols are usually not as stable as keto compounds. Phosphoenol pyruvate, the product of enolase, is unable to tautomerize to the keto form because of the phosphate group. (Recall from organic chemistry that tautomers are compounds that react as if they were made up of two components, differing only in the placement of a substituent, like a hydrogen atom.) Therefore, there is a large negative free energy change associated with release of the phosphate; phosphate release allows the formation of the keto tautomer—that is, of pyruvate. This free energy change is more than enough to phosphorylate ADP to make ATP in the reaction catayzed by **pyruvate kinase:**

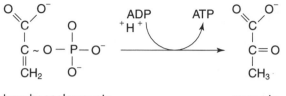

phosphoenolpyruvate                    pyruvate

This reaction, which is highly favored thermodynamically, brings glycolysis into positive energy balance because two ATP bonds are made—one from each of the 3-carbon units from glucose.

The overall reaction of glycolysis is therefore:

glucose + 2 ADP + 2 $P_i$ + 2 NAD $\rightarrow$ 2 pyruvate + 2 ATP + 2 NADH + 2 $H^+$ + 2 $H_2O$

This still leaves one bit of unfinished business. The NAD converted to NADH in the glyceraldehyde-3-phosphate dehydrogenase reaction must be regenerated; otherwise glycolysis could not continue for very many cycles. This regeneration can be done anaerobically, with the extra electrons transferred to pyruvate or another organic compound, or aerobically, with the extra electrons transferred to molecular oxygen, with the generation of more ATP molecules.

**Electron transfer to pyruvate**
The simplest way of regenerating NAD is simply to transfer the electrons to the keto group of pyruvate, yielding lactate, in the reaction catalyzed by **lactate dehydrogenase.** This reaction takes place in animal cells, especially muscle cells, and is carried out by lactic acid bacteria in the fermentation of milk to yogurt.

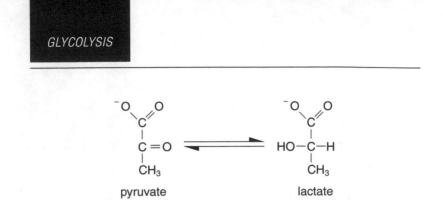

pyruvate                    lactate

The formation of lactate oxidizes the two NADH molecules into NAD; therefore, the glycolytic breakdown of one molecule of glucose becomes:

$$\text{glucose} + 2\ \text{ADP} + 2\ \text{P}_i \rightarrow 2\ \text{lactate} + 2\ \text{ATP} + 2\ \text{H}_2\text{O}$$

**Ethanol**

Ethanol results from the decarboxylation of pyruvate and the reduction of acetaldehyde. Yeasts and other organisms that produce ethanol use a two-step reaction sequence. First, *pyruvate decarboxylase* releases $CO_2$ to make acetaldehyde. Then *alcohol dehydrogenase* transfers a pair of electrons from NADH to the acetaldehyde, resulting in ethanol.

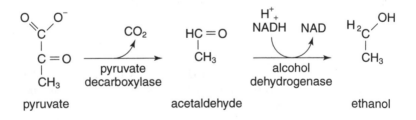

pyruvate                    acetaldehyde                    ethanol

When ethanol is produced, the reaction of glycolysis becomes:

$$\text{glucose} + 2\ \text{ADP} + 2\ \text{P}_i \rightarrow 2\ \text{ethanol} + 2\ \text{ATP} + 2\ \text{H}_2\text{O}$$

The preceding equation explains some traditional winemaking practices. Grapes with the highest sugar content generally make the best wine. On the other hand, unfortified wines have a maximum alcohol content of about 14%, because ethanol inhibits growth and fermentation at that concentration.

The alcohol dehydrogenase reaction occurs in the opposite direction when ethanol is consumed. Alcohol dehydrogenase is found in liver and intestinal tissue. The acetaldehyde produced by liver alcohol dehydrogenase may contribute to short- and long-term alcohol toxicity. Conversely, different levels of intestinal alcohol dehydrogenase may help explain why some individuals show more profound effects after only one or two drinks than others. Apparently, some of the ethanol consumed is metabolized by intestinal alcohol dehydrogenase before it reaches the nervous system.

**Pyruvate to acetyl-Coenzyme A: The entry point into the TCA cycle**

Pyruvate can be oxidatively decarboxylated to form acetyl-Coenzyme A, which is the entry point into the TCA cycle. Louis Pasteur noted in the 1860s that the consumption of glucose by yeast is inhibited by oxygen. This is a regulatory phenomenon, whereby high levels of ATP formed by oxidative metabolism lead to the allosteric inhibition of crucial enzymes in the glycolytic pathway. How does oxidative metabolism form more ATP than fermentation does? Because the carbons from glycolysis are fully oxidized to $CO_2$ through the TCA cycle. The reducing equivalents produced by these oxidations are transferred to molecular oxygen, forming $H_2O$. More free energy is available from the complete oxidation of carbons to $CO_2$ than from the partial oxidations and reductions resulting from anaerobic glycolysis. The biochemistry of this series of reactions is considered in the next chapter.

Glycolysis Regulation

It is a general rule of metabolic regulation that *pathways are regulated at the first committed step*. The committed step is the one after which the substrate has only one way to go. Because glycolytic intermediates feed into several other pathways, the regulation of glycolysis occurs at more than one point. This allows the regulation of several pathways to be coordinated. For example, dihydroxyacetone phosphate is the precursor to the glycerol component of lipids. An animal in a well-fed state synthesizes fat and stores it for energy. Glycerol is needed for formation of triglycerides, even though ATP synthesis is less important. Metabolic control must therefore allow glucose to be converted into triose even though the complete breakdown of the trioses to $CO_2$ need not occur at such a high rate.

The free energy diagram of glycolysis shown in Figure 9-1 points to the three steps where regulation occurs. Remember that for any reaction, the free energy change depends on two factors: the free energy difference between the products and reactants in the standard state and the concentration of the products and reactants. In the figure, the standard free energies and the concentrations were used to compute the total free energy differences between products and reactants at each step. Reactions at equilibrium have a free energy change of zero.

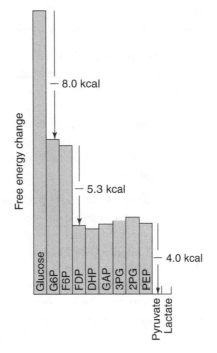

Figure 9-1

## Regulation occurs at the three reactions far from equilibrium

Remember that at equilibrium the rates of forward and reverse reactions are equal. Therefore, the conversion of, for example, 3-phosphoglycerate to glyceraldehyde-3-phosphate occurs rapidly. In contrast, the reactions far from equilibrium, such as the conversion of phosphoenol pyruvate to pyruvate, have rates that are greater in the forward than in the reverse direction. Imagine a series of pools in a fountain. If two pools are at the same level, there is no point in putting a dam between them to control the flow of water. On the other hand, the rate of water flow can be controlled effectively at any point where one pool spills into a lower one. Think of the compounds in the free energy diagram as pools—where does a pool spill into a lower one, offering the possbility of control? At three enzyme-catalyzed reactions:

1. **Glucose-6-phosphate formation.** The entry point of glucose is the formation of glucose-6-phosphate. Hexokinase is feedback-inhibited by its product, so the phosphorylation of glucose is inhibited if there is a buildup of glucose-6-phosphate. In mammalian cells, the breakdown of glycogen is regulated by covalent modification of glycogen phosphorylase (see Chapter 12). This regulation reduces the rate of formation of glucose-6-phosphate.

2. **Fructose-6-phosphate⇨fructose-1,6-bisphosphate.** Glucose-6-phosphate has other metabolic fates than simply leading to pyruvate. For example, it can be used to synthesize ribose for DNA and RNA nucleotides. The most important regulatory step of glycolysis is the phosphofructokinase reaction. Phosphofructokinase is regulated by the **energy charge** of the cell—that is, the fraction of the adenosine nucleotides of the cell that contain high-energy bonds. Energy charge is given by the formula:

$$[ATP] + \tfrac{1}{2}[ADP]/([ATP] + [ADP] + [AMP])$$

The energy charge of a cell can vary from about 0.95 to 0.7. *ATP inhibits* the phosphofructokinase reaction by raising the $K_m$ for fructose-6-phosphate. AMP activates the reaction. Thus, when energy is required, glycolysis is activated. When energy is plentiful, the reaction is slowed down.

Phosphofructokinase is also activated by fructose-2,6-bisphosphate, which is formed from fructose-1-phosphate by a second, separate phosphofructokinase enzyme—phosphofructokinase II (as shown in Figure 9-2). The activity of PFK II is itself regulated by hormone action.

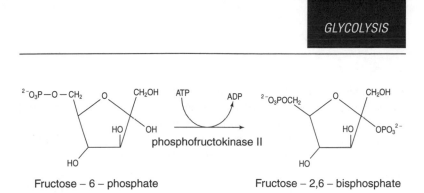

Fructose – 6 – phosphate                Fructose – 2,6 – bisphosphate

Figure 9-2

Fructose-2,6-bisphosphate allosterically activates PFK I by decreasing the $K_m$ for fructose-6-phosphate.

Finally, phosphofructokinase is *inhibited by citrate.* Citrate is the TCA cycle intermediate where 2-carbon units enter the cycle (see Chapter 10). A large number of compounds—for example, fatty acids and amino acids—can be metabolized to TCA cycle intermediates. High concentrations of citrate indicate a plentiful supply of intermediates for energy production; therefore, high activity of the glycolytic pathway is not required.

3.  **Phosphoenol pyruvate → pyruvate.** The third big step in the free-energy diagram is the pyruvate-kinase reaction, where ATP is formed from phosphoenol pyruvate. *ATP inhibits* pyruvate kinase, similar to the inhibition of PFK. Pyruvate kinase is also *inhibited by acetyl-Coenzyme A,* the product of pyruvate metabolism that enters the TCA cycle. Fatty acids also allosterically inhibit pyruvate kinase, serving as an indicator that alternative energy sources are available for the cell.

Pyruvate kinase is also *activated by fructose-1,6-bisphosphate.* Why fructose-1,6-bisphosphate? It is an example of **feed-forward** activation. This glycolytic intermediate is controlled by its own enzyme system. If glycolysis is activated, then the activity of pyruvate kinase must also be increased in order to allow overall carbon flow through the pathway. Feed-forward activation ensures that the enzymes act in concert to the overall goal of energy production.

**Glycolysis produces short but high bursts of energy**

Physiologically, glycolysis produces energy at a high rate but for a short duration. Biopsies of animal muscle indicate two types of tissue; the two types have different metabolic activities. The flight muscles in the breasts of chickens and turkeys, for example, are light, while the leg and other muscles are dark. The color of the dark meat comes from the iron present in the cytochromes involved in oxygen-consuming respiration. In these tissues, metabolism of glucose is largely aerobic. In contrast, flight muscle (a fast-white muscle) contains few mitochondria; glucose is broken down largely by glycolysis. Because only two ATP molecules are produced per glucose consumed by glycolysis, a limited amount of energy is available for muscle activity. The muscle acts quickly, but for only a short time. In contrast, mitochondrial oxidation of glucose in slow-red muscle makes more ATP but the process takes longer. Slow-red muscle doesn't work as quickly as fast-white muscle but can be active for a longer period of time.

Athletes' muscle composition reflects their relative sports. An untrained adult male's leg muscle is about half of each type. Sprinters contain more fast-white muscle, while an elite marathon runner can have as much as 90% slow-red muscle tissue. It is unclear how much of the difference is due to training and how much to heredity.

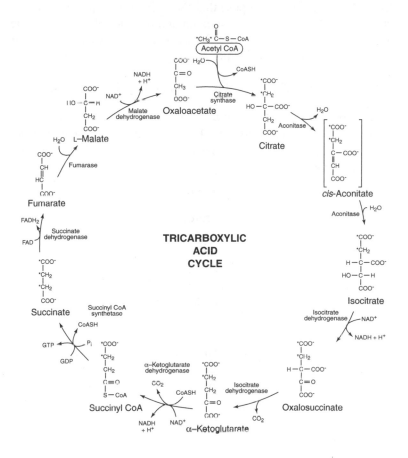

Figure 10-1

The TCA cycle, shown in Figure 10-1, differs from glycolysis (see Chapter 9) in that it has no beginning or end. Whereas adding a given amount of a glycolytic intermediate results in the synthesis of an equimolar amount of pyruvate, the addition of a given amount of an intermediate of the TCA cycle results in a greater than equimolar amount of pyruvate consumed. This **catalytic** behavior of the pathway intermediates was one of the more important pieces of evidence that led Hans Krebs to propose that the oxidation of pyruvate was a *cyclical* pathway. Although the individual molecules of the TCA cycle do not regenerate, each turn of the cycle regenerates an equimolar amount of the acceptor molecule.

The TCA cycle can be thought of as comprising three phases:

1. **Entry phase:** Pyruvate is decarboxylated and the remaining 2-carbon unit is joined to a 4 carbon dicarboxylic acid to make citrate, which is then rearranged into another 6-carbon acid, isocitrate:

$$3 = 2 + 1; 2 + 4 = 6; 6 = 6$$

2. **Oxidative phase:** Isocitrate is decarboxylated, releasing $CO_2$ and reducing equivalents. The product of this reaction is itself decarboxylated to yield a 4-dicarboxylic acid, $CO_2$, and reducing equivalents. The reducing equivalents, mostly as NADH, are then transferred through the cytochromes to an electron acceptor, for example, oxygen:

$$6 = 5 + 1; 5 = 4 + 1$$

3. **Regeneration phase:** The 4-carbon dicarboxylic acid is rearranged to regenerate the acceptor of 2-carbon units. More reducing equivalents are produced:

$$4 = 4$$

## The First Phase of the TCA Cycle

Entry of 2-carbon units is carried out by pyruvate dehydrogenase and citrate synthase in the first phase of the TCA cycle. Pyruvate from glycolysis or other pathways enters the TCA cycle through the action of the **pyruvate dehydrogenase complex,** or **PDC.** PDC is a multienzyme complex that carries out three reactions:

1. **Removal of $CO_2$ from pyruvate.** This reaction is carried out by the pyruvate decarboxylase (E1) component of the complex. Like yeast pyruvate decarboxylase, responsible for the production of acetaldehyde, the enzyme uses a thiamine pyrophosphate cofactor and oxidizes the carboxy group of pyruvate to $CO_2$. Unlike the glycolytic enzyme, acetaldehyde is not released from the enzyme along with $CO_2$. Instead, the acetaldehyde is kept in the enzyme active site, where it is transferred to Coenzyme A.

2. **Transfer of the 2-carbon unit to Coenzyme A.** This reaction is carried out by the dihydrolipamide transacetylase (E2) component of the complex. Lipoic acid is an 8-carbon carboxylic acid with a disulfide bond linking the 6 and 8 carbons:

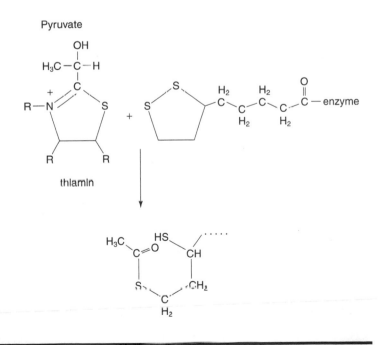

Lipoic acid is bound in an amide linkage with the terminal amino group of a lysine side chain. This long side chain means that the disulfide group of lipoic acid is capable of reaching several parts of the large complex. The disulfide reaches into the adjacent $E_2$ portion of the complex and accepts the 2-carbon unit on one sulfur and a hydrogen atom on the other. Therefore, the oxidized disulfide is reduced, with each sulfur accepting the equivalent of one electron from the pyruvate carboxylase subunit.

The lipoic-acid-bound acetyl group is transferred to another thiol, the end of **Coenzyme A**, a cofactor composed of an ADP nucleotide bound through its phosphates to pantothenic acid, a vitamin, and finally, an amide with mercaptoethylamine. The acetyl group on lipoic acid is transferred to the free thiol (-SH) group of Coenzyme A, leaving the lipoic acid with two thiols:

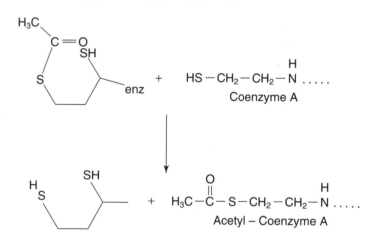

Acetyl-CoA is the substrate for formation of citrate to initiate the TCA cycle.

3. **Regeneration of the disulfide form of lipoic acid and release of electrons from the complex.** This reaction is carried out by the third component of the pyruvate dehydrogenase complex—dihydrolipoamide dehydrogenase ($E_3$). This component contains a tightly bound cofactor—flavin adenine nucleotide, or FAD. FAD can function as a one- or two- electron acceptor. In the reaction catalyzed by $E_3$, FAD accepts two electrons from the reduced lipoic acid, leaving the side chain in a disulfide form. The reduced $FADH_2$ is regenerated by transferring two electrons from $FADH_2$ to NAD (see Figure 10-2).

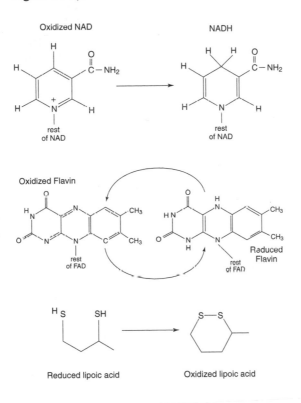

Figure 10-2

In summary, the reactions of the complex are:

- $E_1$: pyruvate + TPP $\rightarrow$ $CO_2$ + hydroxyethyl-TPP
- $E_1$: TPP + pyruvate $\rightleftharpoons$ $CO_2$ + E1: H $\rightleftharpoons$ TPP
- $E_1$ + $E_2$: hydroxyethyl-TPP + lipoic acid $\rightarrow$ acetyl-lipoic acid + TPP
- $E_2$: acetyl-lipoic acid + Coenzyme A $\rightarrow$ acetyl-CoA + $E_2$: lipoic acid$_{reduced}$
- $E_2$: lipoic acid$_{reduced}$ + $E_3$ FAD $\rightarrow$ $E_2$<: lipoic acid + $E_3$: $FADH_2$
- $E_3$: $FADH_2$ + NAD $\rightarrow$ $E_3$: FAD + NADH + $H^+$

Summing up the equations and canceling out the intermediates that appear on both sides of the summed equation yields the overall reaction:

$$\text{pyruvate} + \text{NAD} + \text{CoA} \rightarrow CO_2 + \text{Acetyl-CoA} + \text{NADH} + H^+$$

Acetyl-CoA reacts with a 4-carbon dicarboxylic acid—oxaloacetate—in the second entry reaction of the TCA cycle, which is catalyzed by **citrate synthase.** In organic chemistry terms, the reaction is an **aldol condensation.** The methyl group of acetyl-CoA donates a proton to a base in the active site of the enzyme, leaving it with a negative charge. The carbonyl carbon of oxaloacetate is electron-poor and is thereby available for conjugation with the acetyl group, making citroyl-CoA. Hydrolysis of this intermediate releases free Co-A and citrate (see Figure 10-3).

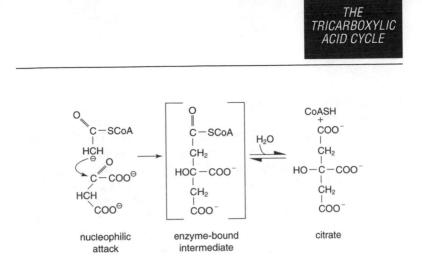

nucleophilic          enzyme-bound          citrate
attack                intermediate

Figure 10-3

Citrate is not a good substrate for decarboxylation. Decarboxylation is usually carried out on alpha-keto acids (like pyruvate, above) or alpha-hydroxy acids. Conversion of citrate into an alpha-hydroxy acid involves a two-step process of water removal (dehydration), making a double bond, and readdition (hydration) of the intermediate—aconitate as Figure 10-4 shows. The enzyme responsible for this isomerization is **aconitase.**

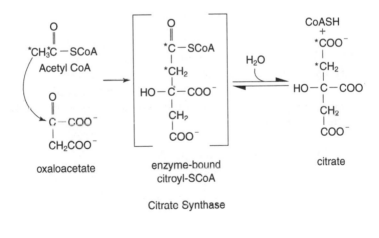

oxaloacetate          enzyme-bound          citrate
                      citroyl-SCoA

Citrate Synthase

Figure 10-4

## Oxidative decarboxylation

Oxidative decarboxylation of isocitrate and alpha-ketoglutarate releases $CO_2$ and reducing equivalents as NADH. The first decarboxylation is a consequence of the oxidation of isocitrate by transfer of two electrons to NAD, catalyzed by **isocitrate dehydrogenase.** Removal of the pair of electrons from the hydroxyl group results in an alpha-keto form of isocitrate, which spontaneously loses $CO_2$ to make alpha-ketoglutarate (see Figure 10-5). This 5-carbon dicarboxylic acid is a participant in numerous metabolic pathways, as it can be easily converted into glutamate, which plays a key role in nitrogen metabolism.

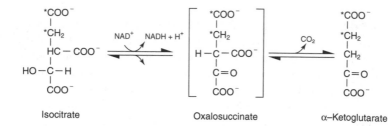

Figure 10-5

Decarboxylation and oxidation of alpha-ketoglutarate is carried out by a large multienzyme complex. Both in the overall reaction it catalyzes and in the cofactors used to carry them out—the alpha-ketoglutarate/dehydrogenase complex (alpha-KGDC)—it is similar to the reaction scheme of the pyruvate dehydrogenase (PDC) complex (see Figure 10-6).

**PDC:** pyruvate + NAD + CoA →
$CO_2$ + Acetyl-CoA + NADH + $H^+$

**alpha-KGDC:** alpha-ketoglutarate + NAD + CoA →
$CO_2$ + succinyl-CoA + NADH + $H^+$

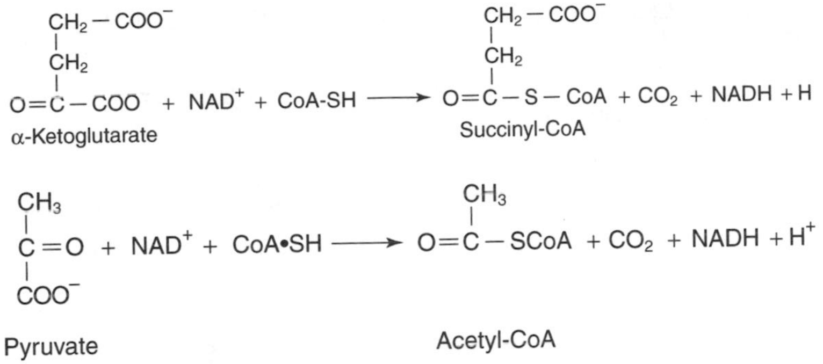

$$\begin{array}{c} CH_2-COO^- \\ | \\ CH_2 \\ | \\ O=C-COO \\ \text{$\alpha$-Ketoglutarate} \end{array} + NAD^+ + CoA\text{-}SH \longrightarrow \begin{array}{c} CH_2-COO^- \\ | \\ CH_2 \\ | \\ O=C-S-CoA \\ \text{Succinyl-CoA} \end{array} + CO_2 + NADH + H$$

$$\begin{array}{c} CH_3 \\ | \\ C=O \\ | \\ COO^- \\ \text{Pyruvate} \end{array} + NAD^+ + CoA{\bullet}SH \longrightarrow \begin{array}{c} CH_3 \\ | \\ O=C-SCoA \\ \text{Acetyl-CoA} \end{array} + CO_2 + NADH + H^+$$

Figure 10-6

Like the pyruvate dehydrogenase complex, the alpha-ketoglutarate dehydrogenase complex has three enzymatic activities, and the same cofactors. As might be expected, the primary sequences of the proteins are highly similar, indicating that they diverged from a common set of ancestral proteins.

The result of this second phase of the TCA cycle is the release of two carbons from citrate. Thus, the equivalent of one mole of pyruvate has been converted to $CO_2$ by this point in the cycle, although the two carbons of acetyl-CoA are still found in succinyl-CoA. The two carbons released as $CO_2$ are derived from the original oxaloacetate involved in the citrate synthase reaction.

### The third phase of the TCA cycle

Succinyl-CoA is hydrolyzed and the 4-carbon dicarboxylic acid is converted back to oxaloacetate in the third phase of the TCA cycle. Succinyl-CoA is a high-energy compound, and its reaction with GDP (in animals) or ADP (in plants and bacteria) and inorganic phosphate leads to the synthesis of the corresponding triphosphate and succinate—a 4-carbon dicarboxylic acid. The substrate-level phosphorylation is catalyzed by **succinyl-CoA synthetase:**

$$\text{Succinyl-CoA} + GDP + P_i \rightleftarrows \text{succinate} + GTP + CoA$$

(Figure 10-7 shows the reaction catalyzed by this enzyme.)

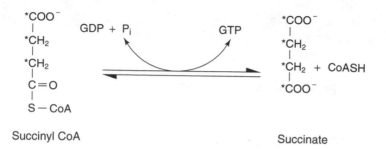

Figure 10-7

Succinate, the 4-carbon saturated precursor to oxaloacetate, then undergoes three successive reactions to regenerate oxaloacetate. The first step is carried out by **succinate dehydrogenase,** which uses FAD as an electron acceptor, as Figure 10-8 shows.

Succinate + FAD $\rightleftharpoons$ fumarate + FADH$_2$

COO$^-$
|
C—H + H$_2$O $\rightleftharpoons$ HO—C—H     $\Delta G^{\circ\prime}$ = –3.8 kJ/mol
||
H—C                H—C—H
|                  |
COO$^-$            COO$^-$

Fumarate           L-Malate

Figure 10-8

Fumarate is the *trans* isomer of the dicarboxylic acid.

Water is added across the double bond in the next step, catalyzed by **fumarase,** to give malic acid, or malate. Finally, **malate dehydrogenase** removes the two hydrogens from the hydroxyl carbon to regenerate the alpha-keto acid, oxaloacetate:

fumarate + $H_2O$ $\rightleftharpoons$ $\times$ malate

malate + NAD $\rightleftharpoons$ $\times$ oxaloacetate + NADH + $H^+$

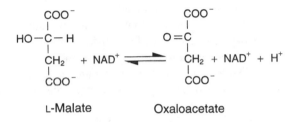

L-Malate          Oxaloacetate

Figure 10-9

## Substrate Availiability, Pyruvate, and the TCA Cycle

The TCA cycle is largely controlled by substrate availability and the entry of pyruvate through the pyruvate dehydrogenase complex. A free energy diagram of the TCA cycle would show a large drop at the three decarboxylation steps. This drop is due to the release of $CO_2$ and the large entropy change associated with this release. Pyruvate dehydrogenase is inhibited by its products — acetyl-CoA and NADH as well as ATP (the end-product of energy metabolism) — and activated by AMP. Isocitrate dehydrogenase and alpha-ketoglutarate dehydrogenase are likewise inhibited by NADH. Isocitrate dehydrogenase is also activated by ADP, and alpha-ketoglutarate dehydrogenase is inhibited by its product, succinyl-CoA. There is only one apparent control point at the 4-carbon level: malate dehydrogenase is inhibited by NADH.

Because the TCA cycle is central to many pathways of metabolism, there must always be a large supply of the intermediates. For example, oxaloacetate is the direct precursor of the amino acid aspartate, with the alpha-keto group being replaced by an amino group. Likewise, alpha-ketoglutarate is the direct precursor to glutamate. These two amino acids are important, not only for protein synthesis, but even more so for maintaining nitrogen balance and eliminating toxic ammonia. Therefore, there are a variety of pathways that serve to regenerate TCA cycle intermediates if the supply falls low. For example, the breakdown of amino acids leads to TCA cycle intermediates. If the supply of intermediates falls low, for example, during even short periods of starvation, muscle can be broken down and the carbon skeletons of the amino acids used to build up the supply of 4-carbon dicarboxylic acids. Oxaloacetate and malate can be synthesized from pyruvate by carboxylation using bicarbonate, the aqueous form of $CO_2$:

$$\text{pyruvate} + HCO_3^- + ATP \rightleftharpoons \text{oxaloacetate} + ADP + P_i + H^+$$

$$\text{pyruvate} + HCO_3^- + NADPH + H^+ \rightleftharpoons \text{malate} + NADP + H_2O$$

Collectively, these reactions are sometimes termed **anapleurotic** reactions. The term is less important than the concept: There needs to be a ready supply of TCA cycle intermediates to keep the system primed for a variety of biochemical reactions.

The reducing equivalents from glycolysis, the Krebs cycle, or other catabolic pathways are carried by coenzymes, particularly NAD, and to some extent FAD. The coenzymes then need to be reoxidized so that the coenzymes can be used again. In anaerobic metabolism, the terminal electron acceptor is a carbon-containing compound, such as pyruvate or acetaldehyde. The Krebs cycle releases carbon as $CO_2$, which can be reduced, but only by a reductant stronger than NADH. In aerobic metabolism, the terminal electron acceptor is oxygen, $O_2$, which is reduced to water:

$$\tfrac{1}{2} O_2 + 2H^+ + 2e^- \rightarrow H_2O$$

This reaction is highly exergonic. The energy of oxidation is the same, whether the reaction occurs in a fire or in a cell. The difference is that the reaction in a cell occurs in a controlled fashion, in small steps. The purpose of the reactions in the *respiratory chain* leading from NADH to oxygen is to conserve the energy of oxidation and convert it to ATP. The reducing equivalents from NADH are transferred through a series of membrane-bound proteins, the **cytochromes.** As the electrons pass through the cytochromes, hydrogen ions, or protons, are released on one side of the membrane, leading to an *electrical potential* across the membrane. The protons flow across the membrane, and the energy associated with this electrical flow is converted to ATP. This overall process by which reducing equivalents are used to make ATP is known as **oxidative phosphorylation.** The process of proton flow leading to ATP synthesis is known as the **chemiosmotic mechanism.**

## Oxidative Phosphorylation

Oxidative phosphorylation occurs on membranes. In bacteria, chemiosmotic ATP synthesis occurs at the cytoplasmic membrane. In plant and animal cells, these reactions occur in the **mitochondrion,** a double-membraned organelle (Figure 11-1). The ancestor of mitochondria was a bacterial cell incorporated into a nucleated cell, which subsequently lost much (although not all) of its DNA. Most mitochondrial proteins are encoded by nuclear DNA. Some respiratory proteins, along with mitochondrial ribosomal RNA and transfer RNAs, are encoded by mitochondrial DNA.

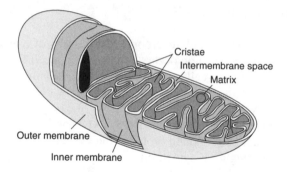

Figure 11-1

The outer membrane of the mitochondrion contains a large number of pores, so that molecules with molecular weights less than 1,000 can pass from the cytoplasm into the **intermembrane space** without any specialized transport mechanisms. This means, for example, that NADH, ADP, and inorganic phosphate can reach the intermembrane space from the cytoplasm while NAD and ATP can reach the cytoplasm. The inner membrane is much less permeable, in part due to the presence of a specialized membrane lipid known as **cardiolipin** (meaning heart lipid—cardiac cells have a large number of mitochondria). The inner membrane of the mitochondrion is highly folded into *cristae,* so that it has an interleaved appearance in the electron microscope.

Inside the inner mitochondrial membrane is the space called the **matrix**. The matrix contains DNA, the apparatus for mitochondrial protein synthesis, and the enzyme systems for the TCA cycle and fatty acid oxidation. (Note that the TCA cycle enzymes are also located in the cytoplasm, because they are involved in many other metabolic transactions in the cell.) Because generation of NADH for oxidative phosphorylation occurs in the matrix, and because the inner membrane is so impermeable, there must be a large number of specific **transport systems** to allow small molecules to reach the site where they are catabolized. Finally, many mitochondrial proteins are made by cytoplasmic ribosomes, encoded by nuclear DNA. This means that there must be protein transporting systems to bring these macromolecules into the matrix.

## The energy of oxidation

The energy of oxidation is given by the redox potential of the reaction. If a piece of copper wire is placed in a solution of zinc sulfate, nothing happens, but if zinc metal is placed in a solution of copper sulfate, the zinc metal is corroded. Simultaneously, the blue color of the copper sulfate solution disappears, and metallic copper is deposited on the surface of the remaining zinc metal:

$$Zn(0) + Cu^{2+} \rightarrow Zn^{2-} + Cu(0)$$

The reaction is the transfer of electrons from zinc to copper ions:

$$Zn(0) \rightarrow Zn^{2+} + 2e^-$$
$$Cu^{2+} + 2e^- \rightarrow Cu(0)$$

The overall reaction is the sum of these two *half-reactions*. The reaction can occur even if the half reactions (that is, the zinc metal and copper ions) are in different containers, as long as the two half reactions are connected electrically, as in Figure 11-2.

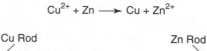

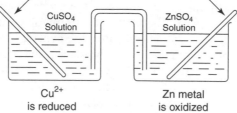

Cu²⁺ is reduced     Zn metal is oxidized

Figure 11-2

The force associated with electron flow is the voltage. In Figure 11-2, the voltage can be measured, either by a voltmeter (similar to a battery tester), or by applying an *opposing voltage* to the wire connecting the zinc and copper electrodes. The amount of voltage required to cancel out the flow of electrons between the electrodes is called the **potential** of the Zn/Cu cell, designated $E$. The voltage of a cell depends on the concentrations (more accurately, the activities) of the ions. Chemists refer reaction energies to a standard state. In this case, the standard state is the one where the ions are present at 1 Molar concentration. The solid metals are always given the activity of 1. In the standard state, the potential of the Zn/Cu cell shown in Fig 11-2 is about 1.1 volts; this means that it can be stopped by application of a direct current from a battery of 1.1 volts opposing the direction of spontaneous electron flow.

The reduction potential is related to the standard free energy change associated with reduction by the equation:

$$\Delta G^{0\prime} = {}^{-}\,n\mathscr{F}E^{0\prime}$$

where $n$ is the number of electrons involved in the reaction, $\mathscr{F}$ is the Faraday of a constant having the value 23.06 kcal Volts$^{-1}$ mol$^{-1}$ (96.5 kJ mol$^{-1}$ V$^{-1}$), and $E^{0\prime}$ is the standard reduction potential for the two half-reactions. Like other biochemical free-energy changes, the standard reduction potential is determined at pH = 7.0.

The standard reduction potential of a reaction is the sum of the two *half-reactions*. For example, the standard free energy change associated with the reduction of pyruvate to lactate:

$$\text{pyruvate} + \text{NADH} + \text{H}^+ \rightleftharpoons \text{lactate} + \text{NAD}$$

is the sum of the two half reactions:

$$\text{pyruvate} + 2e^- + 2\text{H}^+ \rightarrow \text{lactate} \qquad \Delta E^{o'} \; -0.19 \text{ V}$$

$$\text{NAD} + 2e^- + \text{H}^+ \rightarrow \text{NADH} \qquad \Delta E^{o'} = -0.32 \text{ V}$$

NADH is oxidized to NAD, so the second half-reaction must be reversed:

$$\text{NADH} \rightarrow \text{NAD} + \text{H}^+ + 2e^- \qquad \Delta E^{o'} = +0.32 \text{ V}$$

The standard reduction potential of the reaction is therefore the sum of the two half-reactions:

$$\Delta E^{o'} = -0.19 + 0.32 = +0.13$$

This corresponds to the free energy change of:

$$\Delta G^{o'} = -(2)(23.06)(+0.13) = -6 \text{ kcal mol}^{-1} \; (-25.1 \text{ kJ mol}^{-1})$$

**Biochemical reduction and concentration-dependency**
This dependence is given by the **Nernst equation:**

$$E' = E^{o'} + \text{RT}/ n\mathscr{F} \ln(\Pi \; [\text{products}]/ \Pi \; [\text{reactants}])$$

In the preceding reaction:

$$E' = E^{o'} + (2.303\text{RT}/ n\mathscr{F})\log([\text{lactate}][\text{NAD}]/[\text{pyruvate}]$$
$$[\text{NADH}]([\text{H}^+]/10^{-7})$$

This is the same form as the dependence of free energy on the concentration of the reductants and products (Chapter 3). The term $([H^+]/10^{-7})$ accounts for the fact that the standard state for biochemical reactions is at pH equal to 7.0.

**The oxidation of NADH**

The oxidation of NADH by molecular oxygen provides a large amount of free energy for ATP synthesis. The reduction potential of oxygen is +0.82 V; as can be seen in the preceding formulas, the oxidation of NADH has a standard potential of +0.32 V; therefore, the standard free-energy change associated with the transfer of electrons from NADH to oxygen to make water is:

$$\Delta G^{o'} = - n\mathscr{F}E^{o'} = -(2)(96.5)(0.82 + 0.32) = -220 \text{ kJ}$$
(-53 kcal)

Compare this with the standard free-energy change of making an ATP from ADP and inorganic phosphate: 31 kJ mol$^{-1}$ (7.4 kcal mol$^{-1}$). The purpose of the respiratory chain is to harness this large free-energy change to efficiently synthesize ATP. Three ATP molecules are made by the respiratory chain during the transfer of electrons from NADH to $O_2$; this corresponds to an efficiency of about 40%. This efficiency is about the same as that of, for example, a diesel engine. Because over 90% of the adenosine nucleotides in the cell are normally fully converted into ATP (see the description of energy charge in Chapter 7), the concentration of reactants is lower than standard, and the concentrations of product (ATP) are higher than in the standard state. Thus, cells are able to convert electrical potential into chemical energy at high (although not 100%) efficiency.

## The Electron Transport Chain

Electrons flow through the electron transport chain to molecular oxygen; during this flow, protons are moved across the inner membrane from the matrix to the intermembrane space. This model for ATP synthesis is called the **chemiosmotic mechanism,** or Mitchell hypothesis. Peter Mitchell, a British biochemist, essentially by himself and in the face of contrary opinion, proposed that the mechanism for ATP synthesis involved the coupling between chemical energy (ATP) and osmotic potential (a higher concentration of protons in the intermembrane space than in the matrix). The inner membrane of the mitochondrion is tightly packed with cytochromes and proteins capable of undergoing redox changes. There are four major protein-membrane complexes.

### Complex I and Complex II

Complex I and Complex II direct electrons to coenzyme Q. Complex I, also called NADH-coenzyme Q reductase, accepts electrons from NADH. The NADH releases a proton and two electrons. The electrons flow through a flavoprotein containing FMN and an iron-sulfur protein. First, the flavin coenzyme (flavin mononucleotide) and then the iron-sulfur center undergo cycles of reduction and then oxidation, transferring their electrons to a *quinone* molecule, **coenzyme Q** (see Figure 11-3). Complex I is capable of transferring protons from the matrix to the intermembrane space while undergoing these redox cycles. One possible source of the protons is the release of a proton from NADH as it is oxidized to NAD, although this is not the only explanation. Apparently, conformational changes in the proteins of Complex I also are involved in the mechanism of proton translocation during electron transport.

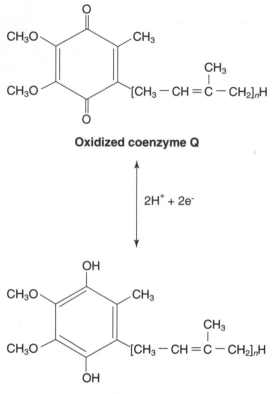

CH₃O and CH₃O substituents on quinone ring with CH₃ group and [CH₃ — CH = C — CH₂]ₙH side chain with CH₃ branch.

**Oxidized coenzyme Q**

$2H^+ + 2e^-$

**Reduced coenzyme Q**

Figure 11-3

Complex II, also known as succinate-coenzyme Q reductase, accepts electrons from **succinate** formed during the TCA cycle (see the previous chapter). Electrons flow from succinate to FAD (the flavin-adenine dinucleotide) coenzyme, through an iron-sulfur protein and a cytochrome $b_{550}$ protein (the number refers to the wavelength where the protein absorbs), and to coenzyme Q. No protons are translocated by Complex II. Because translocated protons are the source of the energy for ATP synthesis, this means that the oxidation of a molecule of $FADH_2$ inherently leads to less ATP synthesized than does the oxidation of a molecule of NADH. This experimental observation also fits with the difference in the standard reduction potentials of the two molecules. The reduction potential of FAD is -0.22 V, as opposed to -0.32 V for NAD.

*Coenzyme Q is capable of accepting either one or two electrons* to form either a *semiquinone* or *hydroquinone* form. Figure 11-4 shows the quinone, semiquinone, and hydroquinone forms of the coenzyme. Coenzyme Q is not bound to a protein; instead it is a mobile electron carrier and can float within the inner membrane, where it can transfer electrons from Complex I and Complex II to Complex III.

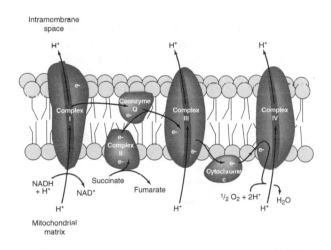

Figure 11-4

Complex III is also known as **coenzyme Q-cytochrome c reductase.** It accepts electrons from reduced coenzyme Q, moves them within the complex through two cytochromes b, an iron-sulfur protein, and cytochrome $c_1$. Electron flow through Complex II transfers proton(s) through the membrane into the intermembrane space. Again, this supplies energy for ATP synthesis. Complex III transfers its electrons to the heme group of a small, mobile electron transport protein, **cytochrome c.**

Cytochrome c transfers its electrons to the final electron transport component, **Complex IV,** or **cytochrome oxidase.** Cytochrome oxidase transfers electrons through a copper-containing protein, cytochrome a, and cytochrome $a_3$, and finally to molecular oxygen. The overall pathway for electron transport is therefore:

$$NADH + H^+ + \tfrac{1}{2}O]_2 \;]+3n \text{ (protons in)} \to NAD + H_2O + 3n$$
(protons out)

or:

$$FADH_2 + \tfrac{1}{2}O_2 + 2n \text{ (protons in)} \to FAD + H_2O + 2n \text{ (protons out)}$$

The number $n$ is a fudge factor to account for the fact that the exact stoichiometry of proton transfer isn't really known. The important point is that more proton transfer occurs from NADH oxidation than from $FADH_2$ oxidation.

## ATP Synthesis

ATP synthesis involves the transfer of electrons from the intermembrane space, through the inner membrane, back to the matrix. The transfer of electrons from the matrix to the intermembrane space leads to a substantial pH difference between the two sides of the membrane (about 1.4 pH units). Mitchell recognized that this represents a large energy difference, because the chemiosmotic potential is actually composed of two components. One component is the difference in hydrogen ion concentration ($pH_{out} - pH_{in}$), symbolized by the term $\Delta pH$. The other component follows from the fact that protons are positively charged, so there is a difference in *electrical potential* symbolized by the term $\Delta\Psi$. The proton gradient results in a state where the intermembrane space is positive and acidic relative to the matrix.

The shorthand for this situation is: **positive out, negative in; acidic out, basic in.** Quantitatively, the energy gradient across the membrane is the sum of the energies due to these two components of the gradient:

$$\Delta M = \Delta \Psi - 2.303 \, \Delta pH / \mathscr{F}$$

The combination of the two components provides sufficient energy for ATP to be made by the multienzyme Complex V of the mitochondrion, more generally known as **ATP synthase.** (See Figure 11-5.)

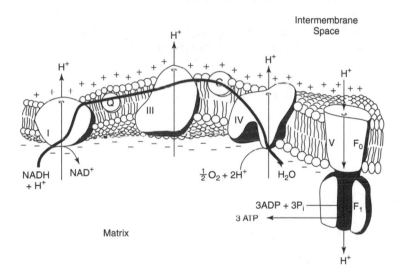

Figure 11-5

ATP synthase contains a membrane-spanning domain, sometimes known as the $F_0$ subunit, and a knobby protrusion that extends into the matrix, the $F_1$ subunit. The mechanism of ATP synthase is not what one would naively predict. The $F_1$ ATP synthase subunit can perform its ligase function (making ATP from ADP and phosphate) without proton flow into the matrix; however, *release of the ATP requires flow of protons through the membrane.*

The existence of ATP synthase implies that electron transport and ATP synthesis are not directly linked. This is borne out by two experimental observations: An artificial proton gradient can lead to ATP synthesis without electron transport, and molecules termed **uncouplers** can carry protons through the membrane, bypassing ATP synthase. In this case, the energy of metabolism is released as heat. One such uncoupler is the compound dinitrophenol, shown in Figure 11-6. Dinitrophenol is a weak acid that is hydrophobic enough to be soluble in the inner membrane. It is protonated in the intermembrane space and deprotonated on the matrix side of the membrane. Because no ATP is made, energy from food is not available for fat synthesis. Indeed, dinitrophenol was used as a diet drug until side effects, including liver toxicity, led to its being withdrawn from the market.

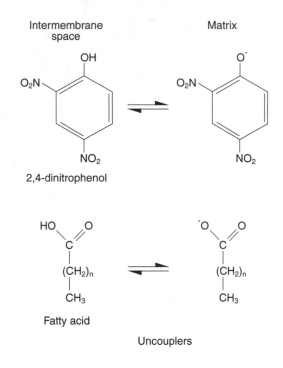

Figure 11-6

Fatty acids are also uncouplers—weak acids that can cross the inner membrane. In human infants, body heat can be generated by so-called brown fat tissues at the base of the neck. The fat appears brown because it contains a high number of mitochondria, and the cytochromes give it a brownish-red appearance. These mitochondria are in a naturally uncoupled state. Fat is oxidized but very little ATP is made; instead, the metabolic energy is converted to heat so that the brain can be kept warm and functioning. Regrettably, the brown fat tissue is lost with age, so adult humans can't burn off their excess calories so easily and naturally.

## Mitochondrial Transport Systems

If the inner membrane is so impermeable, and ATP is made in the matrix side of the membrane, how does it get out into the cell where it's needed? Specific transport systems use either the electrical ($\Delta\Psi$) or acid/base ($\Delta pH$) components of the proton gradient to move substrates in and out of the matrix. (See Figure 11-7.)

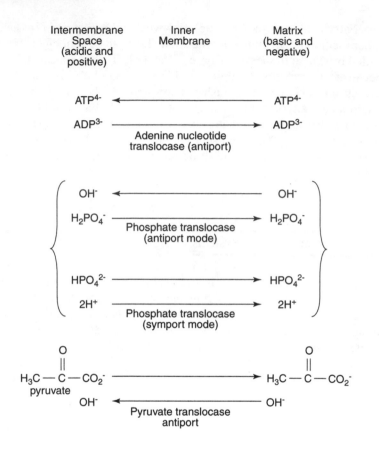

The **adenosine nucleotide transporter** carries out the following exchange reaction:

$$(ADP^{3-})_{out} + (ATP^{4-})_{in} \rightarrow (ATP^{4-})_{out} + (ADP^{3-})_{in}$$

This reaction is powered by the electrochemical component of the mitochondrial gradient, which is positive in the intermembrane space relative to the matrix. The equation, as written, moves a negative charge from the matrix to the intermembrane space, the direction favored by the gradient.

The other substrate for ATP synthase—inorganic phosphate—can come in by either of two modes in response to either component of the gradient. The **phosphate translocase** can operate in an *antiport* mode, transferring hydroxide ion (OH$^-$) out of the matrix in response to the pH component of the gradient:

$$(H_2PO_4^-)_{out} + (OH^-)_{in} \rightarrow (OH^-)_{out} + (H_2PO_4^-)_{in}$$

In this case, there is no net charge flow and the reaction is favored by the fact that the matrix is more basic than the intermembrane space. Alternatively, the phosphate translocase can operate in a *symport* mode:

$$(HPO_4^{2-})_{out} + (H^+)_{out} \rightarrow (H^+)_{in} + (HPO_4^{2-})_{in}$$

In this case, the pH component of the gradient allows the transport of phosphate along with protons. Note that the protons do not go through the ATP synthase in this case.

Carboxylic acids such as pyruvate, succinate, and citrate are transported into the matrix by the **pyruvate transporter,** the **dicarboxylic acid transporter,** and the **tricarboxylic acid transporter,** respectively. Pyruvate transport operates as an antiporter with hydroxide ion. The other transporters are driven by concentration gradients for their substrates. For example, high concentrations of citrate in the matrix lead to export of citrate to the cytoplasm, where it can inhibit phosphofructokinase (see Chapter 9).

---

Energy Yields from Oxidative Phosphorylation

---

Because NADH enters the oxidative phosphorylation at Complex I, three steps of proton translocation result from electron transport, leading to three equivalents of ATP made by ATP synthase. Substrates oxidized by NAD are said to have a **P/O ratio** (phosphates fixed per oxygen atom reduced) of three. Substrates oxidized by FAD, primarily succinate, have a P/O ratio of two. This consideration allows a calculation of the energy fixed from the complete oxidation of a mole of glucose:

- Two ATPs result from glycolysis, with 2 NADH produced by glyceraldehyde $^-3^-$ phosphate dehydrogenase.

- Two turns of the TCA cycle, with NADH produced at the pyruvate dehydrogenase, isocitrate dehydrogenase, alpha-ketoglutarate, and malate dehydrogenase steps and $FADH_2$ produced at the succinate dehydrogenase step and GTP (equivalent to ATP) produced at the succinyl-coenzyme A synthetase step.

In summation:

3 ATPs × 10 NADHs + 2ATPs × 2 $FADH_2$s + 2 ATPs + 2 GTPs =38 ATP high energy bonds per glucose consumed

Comparison of this energy yield with the two ATPs produced by glycolysis alone indicates the enormous competitive advantage allowed by aerobic metabolism in oxygen-rich environments.

Although glucose is the most common sugar, many other carbohydrate compounds are important in cell metabolism. The pathways that break down these sugars yield either glucose or other glycolytic intermediates. Additionally, these pathways can operate in the anabolic direction to transform glycolytic intermediates into other compounds. This chapter considers the degradative pathways first and the biosynthetic pathways secondly.

## The Pentose Phosphate Pathway

It's an unfortunate myth that calories consumed as sugar are better than calories consumed as fat. Both can lead to obesity, if enough are consumed. Foods normally touted as low fat, like fruits, vegetables, and grains, are generally not as calorie-dense as "high-fat" foods, like meat and chocolate candy. Pure carbohydrates yield about 5 kcal of energy per gram and fat about 9 kcal per gram, so the 200 kcal as cocoa (stearic acid) in a small candy bar and the 200 kcal as sugar in a can of soda will contribute equally to obesity. So would the 100 kcal in an apple, except that one tends to eat fewer apples at a sitting. (There's no free lunch—in several senses!) Glucose is converted to pyruvate and then to acetyl-CoA, which is used for fatty acid synthesis. Fatty acids are reduced relative to the acetyl groups, so reducing equivalents (as NADPH) must be provided to the fatty acid synthetase system. The NADPH comes from the direct oxidation of glucose-6-phosphate. Although NAD and NADP differ only by a single phosphate group, their metabolic roles are very different. NAD is kept oxidized so that it is a ready electron acceptor, as in glyceraldehyde-3-phosphate dehydrogenase and the TCA cycle. Most of the NADP pool exists in the reduced form, as NADPH. The NADPH is kept ready to donate electrons in biosynthetic reactions.

The pentose phosphate pathway oxidizes glucose to make NADPH and other carbohydrates for biosynthesis (see Figure 12-1). The major route for reduction of NADP to NADPH is the reaction of glucose-6-phosphate through two successive reactions. In the first, carbon 1 of glucose is oxidized *from an aldol to an ester* form (actually, an internal ester, called a lactone) by glucose-6-phosphate dehydrogenase. In the second reaction, the same carbon is *further oxidized to CO₂* and released, leaving behind a 5-carbon sugar, in a reaction catalyzed by 6-phosphogluconolactone-dehydrogenase. Both reactions *reduce NADP to NADPH.* The 5-carbon residue is **ribulose-5-phosphate.**

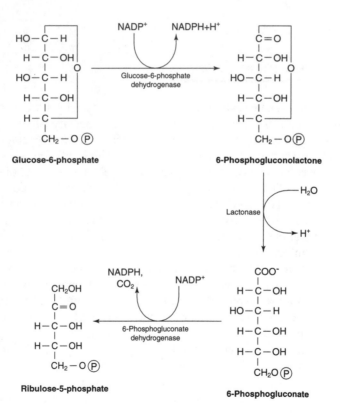

Figure 12-1

These **oxidative reactions** that remove electrons from glucose are a major source of the reducing power for biosynthesis. Accordingly, these enzymes are very active in adipose (fatty) tissue. The oxidation of glucose-6-phosphate to ribulose-5-phosphate and $CO_2$ is also very active in mammalian red blood cells, where the NADPH produced by the reaction is used to keep the glutathione inside the cell in a reduced state. Reduced glutathione helps prevent the oxidation of the iron in hemoglobin from Fe(II) to Fc(III). Hemoglobin containing Fe(III) is not effective in binding $O_2$.

## Ribulose-5-phosphate

Ribulose-5-phosphate has several fates. On one hand, it can be **isomerized** (converted without a change in molecular weight) to ribose-5-phosphate, which is incorporated into nucleotides and deoxynucleotides:

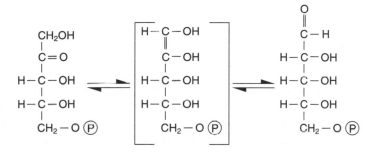

Ribulose-5-phosphate      Enediol Intermediate      Ribose-5-phosphate

Cells that are actively growing need an adequate supply of nucleotides to support RNA and DNA synthesis, and this reaction meets that need.

Alternatively, ribulose-5-phosphate can be converted into another 5-carbon sugar by **epimerization** (change of one stereoisomer into another) into another pentose, **xylulose-5-phosphate.** This reaction is at equilibrium in the cell:

Ribulose-5-phosphate                     Xylulose-5-phosphate

## Converting pentoses to sugars

The pentoses are converted into 6- and 3-carbon sugars. This reaction scheme appears complicated, and it is. The way to decipher it is to remember two key concepts:

1. Either 3-carbon units (one reaction) or 2-carbon units (two reactions) are transferred between acceptor and donor molecules. The enzyme responsible for the 3-carbon transfers is called **transaldolase,** and the enzyme that is responsible for the transfer of 2-carbon units is called **transketolase.**

2. The number of carbons involved in the reactions add up to either ten (two reactions) or nine (one reaction).

The first reaction has the shorthand notation:

$$5 + 5 = 3 + 7$$

which stands for the reaction of ribulose-5-phosphate and xylulose-5-phosphate with transketolase (2-carbon transfer):

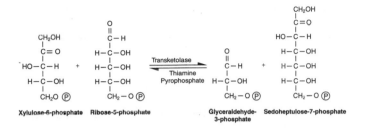

Xylulose-6-phosphate   Ribose-5-phosphate                Glyceraldehyde-   Sedoheptulose-7-phosphate
                                                          3-phosphate

As shown in Figure 12-2, the 7-carbon sugar, sedoheptulose-7-phosphate, and the 3-carbon sugar, glyceraldehyde-3-phosphate, react again, in a reaction catalyzed by transaldolase (3-carbon transfer):

$$7 + 3 = 4 + 6$$

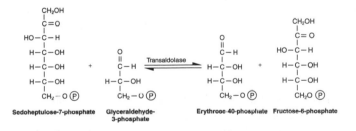

Sedoheptulose-7-phosphate   Glyceraldehyde-        Erythrose-40-phosphate   Fructose-6-phosphate
                            3-phosphate

Figure 12-2

The overall conversion, then, is the conversion of two pentoses into a tetrose (4-carbon) molecule and a hexose. Fructose-6-phosphate, the hexose, is a glycolytic intermediate and can enter that pathway at this stage. As shown in Figure 12-3, the 4-carbon sugar, erythrose-4-phosphate, reacts with a molecule of xylulose-5-phosphate, catalyzed by transketolase (2-carbon transfer):

$$4 + 5 = 6 + 3$$

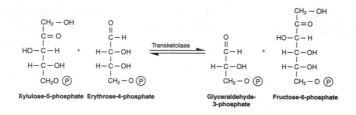

Xylulose-5-phosphate    Erythrose-4-phosphate        Glyceraldehyde-        Fructose-6-phosphate
                                                      3-phosphate

Figure 12-3

The overall reaction scheme of the pentose phosphate pathway is:

$$6 + 6 + 6 = (3 \times 1) + 5 + 5 + 5 \quad \text{Oxidative phase}$$

$$5 + 5 + 5 = 6 + 6 + 3 \quad \text{Sugar interconversion phase}$$

In the sugar interconversion phase, three molecules of ribulose-5-phosphate have thus been converted to two molecules of fructose-6-phosphate and one molecule of glyceraldehyde-3-phosphate. These molecules are glycolytic intermediates and can be converted back into glucose, which can, of course, be used for the synthesis of glycogen.

**Catabolism of other carbohydrates**

The catabolism of other carbohydrates involves their conversion into glycolytic intermediates. Humans encounter a variety of **disaccharides** (two-sugar compounds) in their diet. Glycerol is a product of fat (triglyceride) digestion. **Lactose** (glucosyl-galactose) is predominant in milk, the primary nutrient for mammalian infants. **Mannose** (glucosyl-glucose) and **sucrose** (glucosyl-fructose) are ingested from cereals and sugars. The first step in their utilization is their conversion to monosaccharides by specific hydrolytic enzymes known as glucosidases. A deficiency in these enzymes can cause a variety of gastrointestinal complaints as the unhydrolyzed disaccharides are poorly absorbed in the small intestine. If not absorbed, the carbohydrates pass into the small intestine, where they feed the bacteria there. The bacteria metabolize the sugars, causing diarrhea and flatulence. **Lactase,** the enzyme responsible for lactose hydrolysis, is not synthesized after weaning by most humans. If these individuals consume dairy products, they show symptoms of lactose intolerance. Addition of purified lactase to milk predigests the lactose, often preventing the symptoms.

Before galactose can be metabolized by the glycolytic pathway, it must be converted into glucose-6-phosphate. The first step in the process is the phosphorylation of galactose into galactose-1-phosphate by galactokinase.

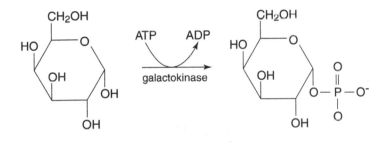

Then galactose-1-phosphate is transferred to a UMP nucleotide by reaction with the sugar nucleotide, **Uridine diphosphate glucose (UDP-glucose).** This reaction liberates glucose-1-phosphate, which is converted into glucose-6-phosphate by phosphoglucomutase (see Figure 12-4). (This enzyme is also important in the breakdown of glycogen.)

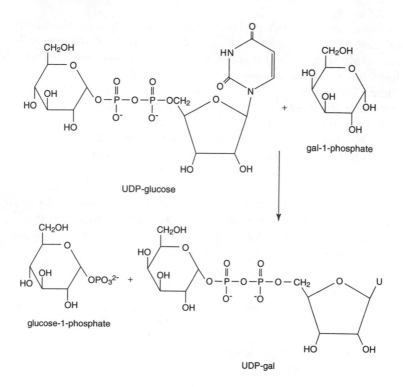

Figure 12-4

UDP-glucose is initially formed by reaction of glucose-1-phosphate with UTP and the release of inorganic pyrophosphate (see Figure 12-5).

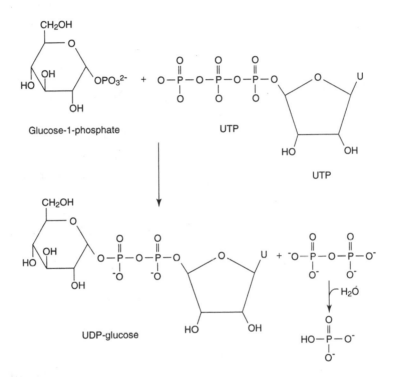

Figure 12-5

Finally, UDP-galactose is epimerized to UDP-glucose by the action of UDP-galactose epimerase (see Figure 12-6). This UDP-glucose can be used in the galactosyltransferase reaction.

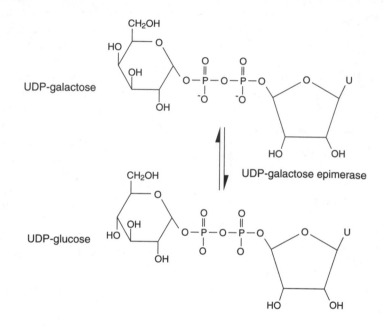

Figure 12-6

This elaborate scheme is probably due to the need to guard against the toxic buildup of galactose-1-phosphate. Humans who lack the enzymes required for galactose epimerization because they have a genetic deficiency of the enzyme suffer from mental retardation and cataracts. In microorganisms, expression of galactokinase in the absence of the epimerase and transferase inhibits cell growth.

## The Gluconeogenic Pathway

The glycolytic pathway can be used for the synthesis of glucose from simpler molecules through gluconeogenesis. **Gluconeogenesis** is the synthesis of glucose from nonsugar sources, especially amino acids and TCA cycle intermediates. Running glycolysis in the synthetic direction requires that there be a way to bypass the three free energy drops in the pathway, that is, the pyruvate kinase, phosphofructokinase, and hexokinase steps (see Chapter 9). The reactions could be run in reverse; however, that wouldn't be very efficient. Because of the relationship between free energy and equilibrium, only a very small amount of, for example, phosphoenolpyruvate can be made by reversing the pyruvate kinase step. The processes that have very little free energy change associated with them operate near equilibrium. Therefore, an addition of enzymatic products, for example, glyceraldehyde-3-phosphate and dihydroxyacetone phosphate, would simply drive the aldolase reaction in the direction of fructose-1,6-bisphosphate. That is, the new equilibrium would favor the synthesis of hexoses.

### Bypassing the pyruvate kinase step

Bypassing the pyruvate kinase step requires oxaloacetate. The oxaloacetate can come from either of two sources. First, various reactions can build up TCA cycle intermediates, among them oxaloacetate. For example, aspartic acid has the same carbon skeleton as oxaloacetate and ammonia can be removed by several means to yield oxaloacetate:

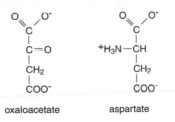

oxaloacetate          aspartate

Alternatively, oxaloacetate can be made from pyruvate by the addition of $CO_2$ in the mitochondrial matrix (see Figure 12-7). This anapleurotic reaction is catalyzed by pyruvate carboxylase. The pyruvate carboxylase reaction consumes an ATP bond and $CO_2$. In eukaryotes, the oxaloacetate thus formed can then be shuttled out of the mitochondria by several pathways (see Chapter 11).

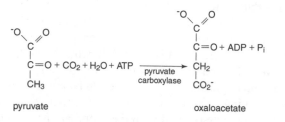

Figure 12-7

Once in the cytosol, the oxaloacetate is decarboxylated and phosphorylated by the enzyme pyruvate carboxykinase (see Figure 12-8). (The nomenclature is confusing; try to remember that pyruvate carboxylase only adds $CO_2$, whereas pyruvate carboxykinase removes $CO_2$ *and* adds phosphate. In the pyruvate carboxykinase reaction, the $CO_2$ added to make oxaloacetate is removed, so the only net reaction in the series is the addition of a phosphate to pyruvate to make phosphoenolpyruvate).

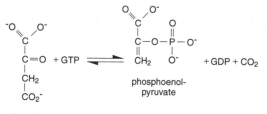

Figure 12-8

However, the two-step reaction consumes two high-energy phosphate bonds to do this task. *A general rule is that biosynthetic pathways use several small, energetically favored steps to bypass a single highly unfavored one in the catabolic pathway.*

### Bypassing the phosphofructokinase and hexokinase steps

The second step that must be bypassed is the phosphofructokinase reaction. The strategy for doing this is simple: A phosphatase hydrolyzes the phosphate at position 1 of fructose bisphosphate, leaving the singly phosphorylated fructose-6-phosphate:

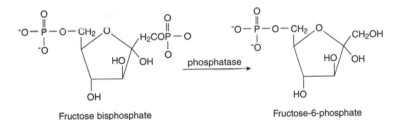

Fructose bisphosphate        Fructose-6-phosphate

If fructose bisphosphate phosphatase and phosphofructokinase were to both operate at high rates in the cell, a great amount of ATP would be broken down to no effect. The two reactions would oppose each other:

Fructose-6-phosphate + ATP →
Fructose-1,6-bisphosphate + ADP

Fructose-1,6-bisphosphate →
Fructose-6-phosphate + phosphate

The sum of these two reactions is simply:

ATP → ADP + phosphate

Fructose bisphosphate phosphatase is regulated by the same allosteric effectors as is phosphofructokinase, except in the opposite manner. For example, phosphatase is activated by fructose-2,6-bisphosphate, whereas phosphofructokinase is inactivated by it. If there were no coordinate regulation of these steps, the net result would be the runaway consumption of ATP in a **futile cycle.** The regulatory mechanism doesn't completely shut down either reaction; rather, it ensures that there is a greater flow of carbon in one direction or the other. The small amount of ATP that is consumed by the futile cycle is the cost associated with the regulation.

The hexokinase step is bypassed in the same manner as is the phosphofructokinase reaction, by a phosphatase that is activated by high concentrations of glucose-6-phosphate (an example of substrate-level control). Note that this is the opposite of the effect of glucose-6-phosphate on the rate of the hexokinase step.

---

## Storage of Glucose in Polymeric Form as Glycogen

The liver secretes glucose into the bloodstream as an essential mechanism to keep blood glucose levels constant. Liver, muscle, and other tissues also store glucose as glycogen, a high-molecular-weight, branched polymer of glucose. Glycogen synthesis begins with glucose-1-phosphate, which can be synthesized from glucose-6-phosphate by the action of phosphoglucomutase (an isomerase). Glucose-1-phosphate is also the product of glycogen breakdown by phosphorylase:

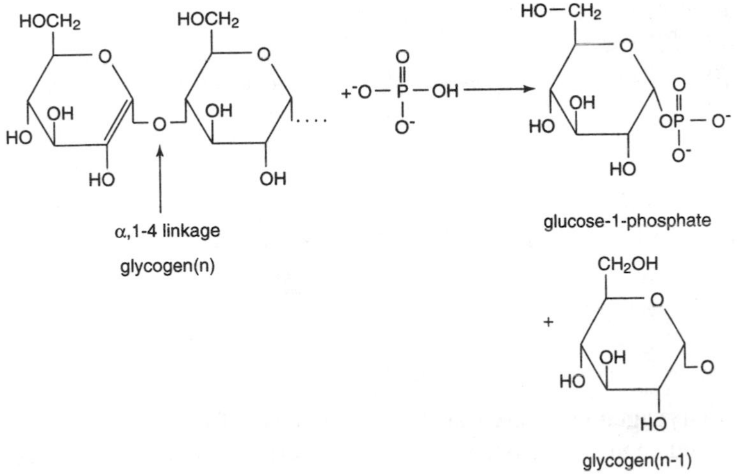

α,1-4 linkage

glycogen(n)

glucose-1-phosphate

glycogen(n-1)

The $K_{eq}$ of the phosphorylase reaction lies in the direction of breakdown. In general, a biochemical pathway can't be used efficiently in both the synthetic and the catabolic direction. This limitation implies that there must be another step in glycogen synthesis that involves the input of extra energy to the reaction. The extra energy is supplied by the formation of the intermediate UDP-glucose. This is the same compound found in galactose metabolism. It is formed along with inorganic pyrophosphate from glucose-1-phosphate and UTP. The inorganic pyrophosphate is then hydrolyzed to two phosphate ions; this step pulls the equilibrium of the reaction in the direction of UDP-glucose synthesis (see Figure 12-9).

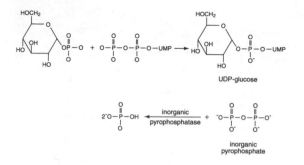

Figure 12-9

**Glycogen synthase** transfers the glucose of UDP-glucose to the nonreducing end (the one with a free Carbon-4 of glucose) of a preexisting glycogen molecule (another enzyme starts the glycogen molecule), making an A, **1-4 linkage** and releasing UDP (see Figure 12-10). This reaction is exergonic, though not as much as the synthesis of UDP-glucose is.

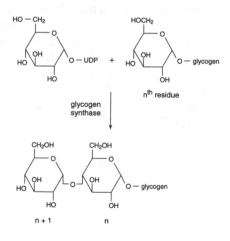

Figure 12-10

Summing up, the synthesis of glycogen from glucose-1-phosphate requires the consumption of a single high-energy phosphate bond and releases pyrophosphate (converted to phosphates) and UDP. Overall, the reaction is:

Glucose-1-phosphate + UTP + $glycogen_n$ →
$glycogen_{n+1}$ + UDP + 2 phosphate

**Glycogen phosphorylase** breaks down glycogen by forming glucose-1-phosphate, in the following reaction:

$glycogen_{n+1}$ + phosphate →
glucose-1-phosphate + $glycogen_n$

This reaction does not require any energy donor. Notice that glycogen breakdown preserves the phosphate of the glucose-1-phosphate that was used for synthesis without the need for a separate phosphorylation step. The sum of the preceding two reactions is simply:

UTP → UDP + phosphate

Since 38 ATPs are made from the oxidative metabolism of a single glucose molecule (see Chapter 11), this minimal energy investment is well worth the advantages of banking the glucose as glycogen.

Glycogen synthase and phosphorylase are reciprocally controlled by hormone-induced protein phosphorylation. One of the most basic physiological reactions in animals is the reaction to danger. The symptoms are probably familiar to anyone who has had to give a public speech: rapid heartbeat, dry mouth, and quivering muscles. They are caused by the hormone epinephrine (adrenaline), which acts to promote the rapid release of glucose from glycogen, thereby providing a rapid supply of energy for "flight or fight."

Epinephrine acts through **cyclic AMP (cAMP),** a "second messenger" molecule.

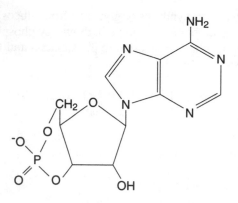

**Cyclic AMP**

The epinephrine receptor causes the synthesis of cyclic AMP, which is an activator of an enzyme, **a protein kinase C** (see Figure 12-11). Protein kinases transfer phosphate from ATP to the hydroxyl group on the side chain of a serine, threonine, or tyrosine. Protein kinase C is a serine-specific kinase. Protein kinase C is a tetramer composed of two regulatory (R) subunits and two catalytic (C) subunits. When it has cAMP bound to it, the R subunit dissociates from the C subunits. The C subunits are now catalytically active.

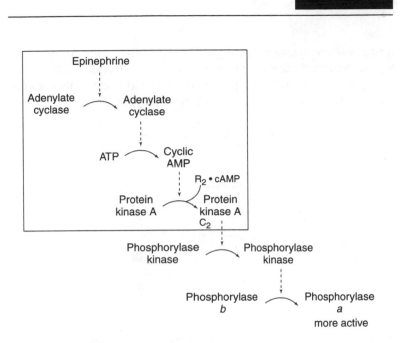

Figure 12-11

Protein kinase C phosphorylates glycogen synthase directly, as well as another protein kinase, **synthase/phosphorylase kinase.** Phosphorylation has different effects on the two enzymes.

Phosphorylation of glycogen synthase, either by protein kinase C or by synthase/phosphorylase kinase, converts it from the more active **I form** (independent of glucose-6-phosphate) to the **D form** (dependent on glucose-6-phosphate). Glycogen synthesis is reduced; although, if glucose-6-phosphate is present in high amounts, the enzyme can still make glycogen.

Phosphorylation of glycogen phosphorylase by synthase/phosphorylase kinase has the opposite effect. The nonphosphorylated form of the enzyme, **phosphorylase b,** is less active than the phosphorylated form, **phosphorylase a** (see Figure 12-12). (Think of *a* for *active* to help remember the direction of regulation.) Phosphorylase a then converts glycogen to glucose-1-phosphate. The end result of this protein phosphorylation cascade is an increased energy supply for activity.

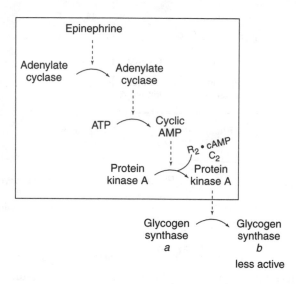

Figure 12-12

**Protein phosphorylation cascades,** like the one discussed above, are a general mechanism of cellular regulation. Protein kinases are involved in the control of metabolism, gene expression, and cell growth, among other processes.

# Notes

# Notes

# Notes

# Notes

# Notes

# Notes